COSMIC ODYSSEY

Cosmic Odyssey

Jean Heidmann
Observatoire de Paris

Translator
Simon Mitton
St Edmund's College University of Cambridge

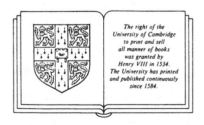

The right of the
University of Cambridge
to print and sell
all manner of books
was granted by
Henry VIII in 1534.
The University has printed
and published continuously
since 1584.

CAMBRIDGE UNIVERSITY PRESS

Cambridge   New York   Port Chester

Melbourne   Sydney

CAMBRIDGE UNIVERSITY PRESS
Cambridge, New York, Melbourne, Madrid, Cape Town, Singapore, São Paulo

Cambridge University Press
The Edinburgh Building, Cambridge CB2 8RU, UK

Published in the United States of America by Cambridge University Press, New York

www.cambridge.org
Information on this title: www.cambridge.org/9780521343770

First published as Jean Heidmann, *L'Odyssée Cosmique* by
Editions Denoël, Paris, and © Editions Denoël 1986.

This English language edition © Cambridge University Press 1989

English language edition first published 1989

This digitally printed version 2008

A catalogue record for this publication is available from the British Library

ISBN 978-0-521-34377-0 hardback
ISBN 978-0-521-34851-5 paperback

Contents

Preface

The universe is a huge place. Three objects in the night sky, visible to the naked eye, can give us some impression of the dizzying depths of space. These three objects, the Moon, the Pole Star and the Andromeda galaxy belong respectively to the planetary, stellar and extragalactic domains. Light, travelling at 300,000 kilometres (186,000 miles) per second, takes one and a quarter seconds to reach us from the Moon, six hundred years from the Pole Star, and two million years to journey from the Andromeda Galaxy.

The universe is also ancient. Its past history is a series of overlapping epochs. Just a few thousand years encompass the historic past, and a few million take us back to the dawn of prehistory. Geologic history extends a few billion years into the past, whereas the cosmological history of the universe, takes us back fifteen billion years to the Big Bang itself.

The universe is full of delights. There are spiral galaxies and gaseous nebulas, faintly glowing mists set against the backdrop of deep space, multiple stars spewing out fantastic arcs of matter, and the fabulous landscapes of planets and their satellites. Humans have walked on the nearest object, the Moon. It always fills me with amazement when I see it in the early evening, between the first quarter and Full Moon, dominating the clear sky, as dusk begins. Reflected sunlight thus impressively shows us that the Moon is a globe. We can see dark plains splashed across the Moon's face, and at the jagged left hand boundary of the gibbous Moon, mountains are just discernible. And we can add to this beauty the astonishing knowledge that the mass of this globe has described a precise and regular orbit around the Earth for four billion years.

The universe is full of mystery. Why does another world accompany our own in its journey through space? What is the purpose of these worlds lost in immense, perhaps infinite, space? Why is space curved and linked to time in the four dimensions of spacetime? What of life, intelligence, and self-awareness, the consequence of life here on Earth? And is there perhaps extraterrestrial intelligence, on other worlds, in a more highly evolved form than here...

The universe is complex. Ever more sophisticated instruments reveal complicated physical processes, bringing renewed challenges for ever more elaborate theories: relativity, quantum mechanics, and grand unification.

The universe is our home. Our destiny and its fate are linked in a fundamental symbiosis. We are dependent on physical processes on scales running from organic molecules through to clusters of galaxies. We depend right now on the rules governing those processes, and from the remote past on the evolution they have experienced. We are even linked to the sub-structure of the elementary particles of matter, which in the last analysis govern physical processes. What are we? We are like the foam on a raging sea.

As humans we want to know where we are and where we are going. We want to understand the destiny that awaits the human race. That destiny intersects with that of the cosmos in which we are immersed, body and soul, and of which we are made. Humans want to understand these things in order to experience the beauty and joy of life to the fullest, and to benefit from the highest intellectual achievements.

The purpose of this book is to give a description of the fabulous richness of the cosmos, and to give as clear a picture as possible of the real nature of our incredible universe. After a first look at the starry night sky, I review the enigmas posed since ancient times by the universe. Then I give a broad brush view of the universe as we understand it today. Following this, a trio of chapters take us to ultimate questions about its nature; we explore in turn the relativistic universe, the quantum universe and the inflationary universe.

Finally I return to questions that touch on our own presence in the universe.

1

The starry night

Night, our window on the universe...

The night is our window on the universe. In the daytime the blueness of the sky prevents us from seeing into space. The blue light is caused when the intense sunlight is scattered by oxygen molecules in the upper atmosphere. Beneath this atmospheric layer, which is a few tens of miles deep, the situation is somewhat like looking through net curtains: if a searchlight were illuminating the fabric, it would be impossible to see anything apart from the searchlight itself. By day we can see only the Sun and the Moon (if it has risen), which is so much fainter by comparison that many people think incorrectly that it is invisible during the day. But when we switch off the searchlight, the surrounding landscape can be discerned: the stars, the planets...

If the Earth did not have an atmosphere, the stars would be visible in broad daylight. The dozen Apollo astronauts who landed on the Moon experienced just such a spectacle, despite the dazzle of sunlight. However, things could be worse. The planet Venus is shrouded in a dense atmosphere and the surface pressure is one hundred times that of the Earth's atmosphere. This blanket is so thick that not a single star is visible at night; even in daytime the Sun itself is invisible and only a feeble glimmer of light reaches the surface. Dense fog or oppressive thunderclouds perhaps give us a vague impression of what the sky must be like on Venus.

What notion of the universe might Venusians have? There would be no stars, planets, comets or nebulas. Seeing only the solid ground, which would seem as motionless as the Earth does to us, and noting the alternation of night and day, they, like us, would come to explore their world, and conclude that it is an isolated sphere in empty space. If they knew a little physics, then they could carry out the Foucault pendulum experiment or use other devices with gyrocompasses in order to conclude that their world rotates nearly in synchronism with the day-night alternation. They would deduce that an undetected light makes an orbit of 225 days' duration around their own world, which in turn rotates on its axis in 243 days. After discovering the laws of gravitation and working out theoretically the properties of the Venusian atmosphere, they could next calculate the minimum distance to the unseen source of light and calculate a minimum value for its rate of energy production. If they then turned to nuclear physics in order to explain this minimum energy requirement, they would arrive at a notion of the Sun that would be more or less correct. But that's all; no further progress could be made. As far as Venusians are concerned, the universe consists entirely of their world orbiting round the Sun. The lack of any precise reference point in space would prevent them from measuring the small perturbations due to the massive planet Jupiter. Of course, by mastering radio astronomy they could see beyond their atmosphere, and by constructing space rockets they could get above it altogether. With that sort of progress everything could in principle be discovered.

So much for Venusians; but what of Earthlings? Are we not to a lesser degree in almost the same situation despite our fine nights? Even though mankind, and animals, have been able to see the stars all down through the ages, their distances, and therefore their real nature, have only been known about for a century. Stars are just like our Sun but they are much further away. Light takes eight minutes to travel from the Sun but it takes four years to get here from the nearest star (situated, to use astronomical jargon, four light years distant). In the realms of the billions of stars that form our Galaxy, the most distant are 100,000 light years away.

So far as planets are concerned, our knowledge is also relatively new. In ancient times, Greek philosophers, Chinese astronomers, and Pre-Columbian watchers of the skies all noticed that among the stars there were five points of light that moved across the sky in an exceedingly complicated fashion. It required precise calculation by Copernicus and Kepler, together with the observations of Galileo, in the sixteenth and early seventeenth centuries to reveal the true nature of these five wanderers, known to us as Mercury, Venus, Mars, Jupiter and Saturn. All are worlds similar to our own, with diameters ranging from 3750 to 85,000 miles (5000 - 140,000 km). They orbit the Sun at between 38 million and more than one billion miles (60 million to one and a half billion kilometres), with periods of 88 days (Mercury) to 29 years (Saturn). In the last twenty years, orbiting satellites and deep space probes have visited these five planets, as well as Uranus, and they have spectacularly expanded our knowledge of the solar system. With the discovery of around sixty natural satellites more than 150 miles (200 km) in diameter we can say a new frontier of discovery has opened.

At the dawn of the twentieth century, the known universe extended no further than the starry realms. The solar system, extending for a few light hours, is centred on an average star, just one of hundreds of billions disposed in a gigantic pancake, the visible evidence of which is the Milky Way, that misty band of light encircling the night sky. This band is due, quite simply, to the super-position of countless faint stars, individually invisible to the naked eye, up to 100,000 light years away, and distributed in the plane of the Galaxy. This is the apparent limit of the visible universe, isolated in infinite space.

However, in the year 964, the astronomer Abd al-Rahman al-Sûfi reported the presence on the celestial sphere of the faint light of another galaxy, in the constellation Andromeda. Actually it was only in 1923 that Edwin Hubble stumbled across its real nature: a galaxy like our own lost in the vastness of extragalactic space. To make this discovery he used the largest telescope available at the time. Since then, the deepest surveys have netted a hundred billion galaxies, stretching to billions of light years at the limit of present techniques. In this century, the discovery of the extragalactic domain has vastly extended the scope of the cosmos. Compared with the volume attributed to the starry universe in the nineteenth century, the extragalactic universe is a million billion times larger...

Mankind is even today in a situation a bit like the imaginary Venusians. Gazing through the window of the night onto their universe, people see the stars, but they know hardly anything more than Venusians might beneath their cloudy skies. That's just the fate of people who live in large cities. How many city-dwellers, closing their windows in the evening, entertain themselves with a vision of the universe, indulge in flights of fancy and contemplate the infinity of the cosmos? Who reflects for a moment on those primordial questions about our place in the cosmos and our destiny in this elaborate scenario? In great cities the night is rare indeed when the Pole Star is not blotted out by atmospheric pollution illuminated by street lighting and garish neon signs. That same Pole Star, reminding us all the time that the Earth, with its towns and cities, its mountains and great oceans, its billions of inhabitants turns once every 24 hours on an axis directed towards it.

City dwellers only experience the starry skies when camping in the mountains, on board a cruise liner, or more readily in the back yard of a home in the country...without television! Nevertheless, these phenomena can be created in the city, in a planetarium. A planetarium is a room in which the celestial phenomena can be projected onto the underside of a hemispherical ceiling, which is painted white. In the centre of the room a planetarium projector throws thousands of points of light onto the dome, creating the illusion of a starry sky.

The visitor to a planetarium relaxes in a reclining chair, which gives a panoramic view of the bowl of the sky. Little by little the ambient lighting is dimmed and the eyes become accustomed to the darkness. The profiles of familiar buildings in the city can be seen on the horizon while appropriate background music sets the scene for the nightly drama. The effect is so realistic as to send a shiver down your spine as the last rays of the setting Sun flicker through the room. Then the stars appear: the brightest first of all, and then the fainter ones progressively as the sky darkens. It is a very dark night with 2500 starry pinpoints of light; we pick out the Great Bear, or Big Dipper, and the misty band of the Milky Way forms an immense arc across the heavens.

The planetarium projector speeds up the natural movements of the stars, and makes it easier to understand their motions. Thus the daily procession of the stars, from their rising in the east to their setting in the west, is simulated in just a minute or so. This fantastic wheeling of the stars also shows that the Great Bear turns around a fixed point, marked by the Pole Star.

The big surprise, on our first session in a planetarium, is learning that most of the sights on a good night are themselves illusions! Take, for example, the irresistible impression that all the stars are at the same distance, just like points of light on a celestial sphere supported by the distant horizon, for all the world as if painted on the underside of a big umbrella. We think that way because the notion of a celestial sphere was ingrained into the human mind in the remote past. In reality, of course, the stars are distributed through space at vastly different distances.

Then there is another illusion: in the course of the night the celestial sphere turns like a rigid body from east to west, as if the surface of the Earth is fixed. This daily motion is a consequence of the Earth's 24-hour rotation relative to the distant stars.

Later, in a typical planetarium showing, the phenomenon of the seasons is shown by speeding up the annual motion of the Sun through the stars on the celestial sphere. The

Sun travels along an inclined great circle, known as the ecliptic. In reality, it is the Earth travelling round its annual orbit of radius 93 million miles (150 million km) that makes it *seem* as if the Sun is appearing in a succession of different points in the sky.

The greatest illusion is found in the motions of the planets. They behave so capriciously and erratically that people have given them astrological significance. The first rational explanations of these motions were extremely complicated and the correct solution was not available until the sixteenth century: the observed planetary paths are a combination of the relative motions of the Earth and the individual planets around the Sun.

Many other illusions have misled mankind: the constellations, those wonderful geometrical figures marked out by the brightest stars, have no special significance for they are, in fact, due to the chance superpositions of nearby and remote stars. Stars may seem point like, but they are really enormous; some are a thousand times the diameter of our Sun and could swallow the orbit of Saturn. Their very large distances from the Earth make them appear rather faint. Then there are the planets; they do not shine with their own light but merely reflect the glory of the Sun. Another example is the misty smudge of light visible in the constellation Andromeda on a very dark night, which is not a cloud of gas, but rather a galaxy like our own, composed of the light of billions of stars, seemingly merged into one entity by the great distance.

Night is our miraculous window on the universe. All the same, an immense period of time passed before mankind came realise the true nature of the universe. And even today each new advance brings with it further surprises.

...and source of our questions

You the reader, me, our ancestors and our descendants are part of the microstructure of the universe, like hoar frost on an icy window. Condensation can form there, or sunlight might heat it, or the glass could shatter into fragments, in which cases the fine patterns would either grow, melt, or vanish for ever. Imagine then what would happen if the Earth's atmosphere became more opaque, or the Sun expanded, or, more drastically if material could disappear, then everything would be totally different for us too. Our destiny is therefore intertwined with the ultimate fate of the universe, which interests us deeply.

This is why human beings have devoted an appreciable fraction of available intellectual effort to this problem for the several thousand years since they became conscious of the cosmos. The philosophers of ancient Greece were probably the first to get to the heart of the matter. They hatched their ideas in a cultural context very different from our own and their thoughts became more or less forgotten, but, in the last millennium, they have come to be accepted by the mass of humanity. Throughout history and among many different civilisations an enormous variety of theories have emerged on the nature and fate of the universe. So, it might appear rather presumptuous for us to select one particular picture and say that it is the true model of the universe. Some notions, barely a century old, are already considered laughable, because they only had an impact in the context of their own times. Yet it is a feature of the scientific method that it causes such obsolescence.

The fundamental differences between the very foundations of Newtonian gravitation, on the one hand, and Einstein's formulation, on the other hand, call into question the presumption that progress is gradual. Each time a new point of view haas been proposed it has been evaluated by a simple overarching criterion: to give a more elegant account of the characteristics of the universe. Thus, Einstein moved beyond all the phenomena explained by Newton and was also able to account for tiny anomalies in the orbit of Mercury, which Newton's theory did not include. But this is not to devalue the work of Copernicus and Galileo, each of whom made solid progress. Among those who would need no persuading of this are the 11 men who, in 1984, were treated to the luxury of making numerous orbits around the Earth by means of the Space Shuttle and the Soyuz-Salyut mission and seeing the reality of our planet from a new viewpoint.

These two examples, on gravitation and on the nature of the Earth, illustrate the differing degrees of confidence that scientists assign to their own work. If they are prepared to chance their arm in the case of the Earth (and in the past some were unfortunately burned at the stake for their beliefs), they won't do so in the case of Einstein's theory of gravitation. They sniff at the possibility of further progress, maybe not too far away, because of some unsatisfactory features of the theory that are not clearly formulated; the most intriguing aspect of these radical views is the extraordinary perspective they give on the ultimate fate of the universe.

In this book I am going to be honest. The level of certainty or the degree of doubt about the things I describe will be clearly established so that readers can find their way through the maze of evidence. Except when my enthusiasm is swept along by fantastic discoveries, I tend to be too conservative. finding the right balance between enthusiasm and prudence would be a small reward for my effort in writing this book!

I want to make a clear statement on another point: astrology. But, before I do so, I want to recall for the first time an impressive sight, not visible to city-dwellers right now. It struck our ancestors in their time as a noble star and left me 'starstruck' much later, in the spring of 1984. What could be more relaxing than a beautiful starry night, with its delicate tracery of thousands of stars, and the constellations looking near enough to reach out and touch? How reassuring it is to see the nightly parade of the stars, that slow but majestic progression from east to west — yesterday, today and tomorrow, recording the march of the seasons and every year of one's life. Is that familiar clockwork in the sky paralleling our own life? In this idyllic sky I saw the red planet Mars, a body unlike other objects in the sky, capricious in its motion and appearance. It moves among the fixed stars like any other planet, but the motion is noticeably erratic: it progresses, turns back, and loops the loop. Generally speaking, it is no more prominent than ordinary stars, but in some years it is particularly bright for several weeks and can be brighter than any star. To see it shining powerfully through the night with its steady red glow is indeed impressive. It's pretty easy to feel a sense of menacing implications. Then, once again, Mars becomes inconspicuous and continues its hesitant and tangled path across the sky.

On reflection, it really isn't too surprising that some of the ancient civilisations found such phenomena so striking that they believed the stars could have a direct influence on the affairs of mankind. From such origins astrology flourished. Eventually, the erratic orbit was explained rationally. The distance to Mars has been measured directly by the Viking landers which arrived in 1976, and it is so large as to make a nonsense of the hypothesis that there is a mysterious influence on humans.

So, in this book I will have nothing to say in favour of astrology . The only point I will concede is that when the Sun goes down we sleep also — except for astronomers! Apart from that whimsical observation, note that in fact the Sun exerts a powerful influence on us through the phenomena of day and night and the changing seasons; there will also be catastrophic consequences from any increase in its energy production. Likewise the Moon, our nearest neighbour, affects us through the changing tides and the poetic influence of a beautiful moonlit night. But there are no distinctly astrological consequences. And that is even more true of Mars and the other more distant planets. Are we being deceived? I don't think so. The truth can be even stranger than fiction. Here is just one example — dating from 1983 — of a physical influence that the planets can exert.

The gravitational pull of the Moon on the oceans causes tides. The liquid mass, sloshing across coastal plains and the ocean floor, causes tidal friction, which slows the Earth's rotation. About 400 million years ago the day length was 22 hours. As the Earth's rotation slows, there is a corresponding increase in the distance of the Moon, just through the laws of motion. Effectively the braking of the Earth's spin is transferring momentum to the Moon and sending it into a more distant orbit. This effect has actually been measured.

Laser reflectors left on the Moon during the Apollo mission of the 1970s have shown an increase in the Moon's distance of several centimetres per year.

There is a further phenomenon. The Earth's rotation axis is inclined to its orbital plane round the Sun and relative to the plane of the Moon's orbit round the Earth. This axis slowly precesses, like a spinning top, and a complete cycle takes 26,000 years under present circumstances. However, as the lunar orbit grows this precession period changes as well. As luck would have it, the precession of the Earth's axis will become synchronised with the tiny gravitational force between the Earth and Saturn in one and a half billion years time. Then resonance will progressively alter the motion of the axis. The obvious example of this phenomenon is a child's swing: by giving quite a small push at the right moment in each swing you can make the child go pretty high. On the cosmic swings and roundabouts, the Earth's axis gets twisted over from its present value of 23° inclination to one of 60°. This causes global climatic change, so the weather would become like that in the polar regions today. The polar circles of latitude today extend to Iceland and the Antarctic. At 60° inclination, they would spread to North Africa, the southern United States, and Australia, totally changing the Earth's climate. And as if that isn't enough, the axis will not be stable, but will oscillate around 60° with a period of 100,000 years, triggering such rapid climatic changes that it is hard to see how life on Earth will be able to adapt. That's what I call the influence of the stars, but it isn't astrology.

2

The universe, a thousand-year enigma

Some Roman cosmology...

I'm going to use a wonderful text, written in Rome at the end of
the reign of Emperor Augustus by the poet Marcus Manilius,
to summarise the thousand-year enigmas of the cosmos. The
extracts are taken from his poem on astronomy. This is more
than two thousand years old and lay forgotten until the tenth
century. The first French translation by Pingré was published
in Paris in 1786. I know all this not because I am a historian,
but because I stumbled across the book by sheer chance on the
stand of a bookseller by the Seine. With my love of astronomy,
the Seine, and old books I just could not resist such a find. It was
my first introduction to the works of Manilius.

Manilius himself was probably not an astronomer.
Instead he drew his knowledge from a variety of Greek and
Roman authors. The special fascination of his work is that he
gives an extensive review of astronomy, and this summary
was made at a pivotal moment during the evolution of western
philosophy. Astronomy flourished in the millennium before
Manilius: the diameter of the Earth had been determined, and
the idea that the Earth was isolated in space was taken

seriously. In the next thousand years the subject was almost destroyed. Manilius' poem suffered the deleterious effects of this evolution of thought, because the greater part of the text is pure astrological fakery. This, however, is preceded by sections on the origin of the world and the ideas of leading philosophers about it, on the shape and contents of the Earth, and the heavens and the stars, and these descriptions are worthy of the time of Augustus. Below I have selected some extracts and interleaved them with my own comments; these introduce the topics in this book, rather than the work of Manilius. For the same reason, I give a literal rather than poetic rendering, which is sufficient for us to appreciate the classical breadth of vision, already more than two millennia old, on this great subject.

> *I sing of the grandeur of the heavens and I shall bring to Earth knowledge of the laws governing its fate. first I shall describe Nature's rich tapestry...*

In my book I will follow this pattern by giving a quick tour of the universe and its contents. Our first glance through the window of the night was just a trailer. We shall learn of cosmology, galaxies, stars, planets, quasars, matter and antimatter, quarks, leptons and baryons, magnetic monopoles, and even black holes: stars with astonishing and mysterious properties. I shall weave into the story other topics, a little superficially perhaps because they are outside my own specialisation: organic molecules, pre-biological molecules, the primordial soup, and early Man and Australopithecus, all of which were of fundamental importance in the past for determining the situation in which we find ourselves right now in the solar system.

> *...and the general arrangement of the universe.*

Next we shall be concerned with spacetime, the curvature of space, the contraction of time, relativistic models of the universe, the expansion of the universe, the evolution of life and of planets or other systems where life might arise, the synthesis of elements inside stars and in the first quarter of an hour of the existence of the universe, the appearance of matter right after the 'zero' of the Big Bang, and the formation of molecules in interstellar space.

> *Does the world recognise only the principle of its existence; could it only be itself; has it always existed and will it exist indefinitely; what was its origin; will it ever draw to a close? How has chaos intermingled the primordial elements without order; when the world was filled with dazzling light, darkness was forced back to the abyss.*

Even today these are still live and poignant questions. At present the Big Bang theory is centre stage. However, I am going to throw in some problems. These have emerged in the last twenty years on account of the almost perfect homogeneity of the universe on the grand scale, as revealed by the discovery of the 3°K cosmic microwave background. I shall add the very promising and extraordinary solution furnished by the theory of the inflationary universe. This novel theory takes us right back to the first instant of our universe and, at the same time, will involve us in the arcane microscopic world of elementary particles. I will also touch on a question that can now be posed, rather hesitantly perhaps, in scientific terms: what preceded the Big Bang?

> *Perhaps the world was created in fire, and the stars, those eyes of Nature, have their existence in the flames that dwell throughout the system, forged by the thunderbolts of the skies.*

Here Manilius recalls the first primordial element postulated by the philosophers of his own times. For us, we would think of the inferno of the Big Bang, the primordial first Light that brought the universe into being, more awe-inspiring than any roll of thunder.

*Water must be the universal facilitator, for without it
matter would always be at rest. And how was it created
in fire, by which it is destroyed?*

Water is the second element of the ancients. It plays a
crucial role in the origin of life, and therefore our own
existence. It is not common in the solar system, the exception
being the Earth, which is the only planet to have abundant
supplies. That terrible element fire, reminds us here of the
expanding Sun which will destroy the Earth in ten billion
year's time, and of the final contraction of space, the reverse of
the Big Bang, in which everything will disappear.

*Neither Earth, nor fire, nor Air nor Water came from
each other; these four elements must be the godhead.*

Here Manilius states the problem of the origin of
everything. Today we wonder: was it a quantum fluctuation in
the vacuum of space that created the matter of our universe
and of countless parallel universes not connected to our own?
That's just one possibility given in the theory of the inflationary
universe. There is another answer to the deepest questions, and
it comes from the anthropic principle. This enables us to
understand how, against all the odds, the universe plays a
game of chance from which humanity emerges at the final
count.

*How was the universe formed, and who are the
creators of all that exists, who only allow us to know
what has gone before?*

I am reminded here of recent theories in which the
universe appears by chance and vanishes into nothing...

*How has everything been arranged in such a way that
cold combines with heat, dryness with humidity, solids
with liquids; always opposed, always working together;*

they are found melded together, capable of growth, and having the potential to produce everything that exists.

This is the problem of the existence of physical laws in nature, the origin of which seems to be even more mysterious than the origin of matter itself. It relates to the physics governing the interactions of quarks, gluons, and bosons, which called 'elementary particles', and are the building blocks of the atoms and molecules that make up the tangible cosmos on which everything hinges.

These various opinions will always be demolished; there will always be uncertainty about the origin of the world; the reason for its existence is concealed from us; it is above the genius of men and gods.

I take my hat off to such a frank expression of modesty, which should not constrain our own investigations with due respect. The great strides made since the time of Manilius give us hope of new progress in the future.

But however obscure the origin may have been, we can at least agree something about the arrangement of its various parts: they are all placed in a fixed order.

Effectively, Manilius recognised that there are accepted facts. Unhappily, two thousand years later, some are no longer acknowledged as such. And it will be the same for our discoveries in future centuries.

Winged fire, most subtle element, reached aloft for the ethereal level. Established in the starry heavens it forms a rampart of flames marking the boundary of the natural world.

Thermonuclear energy makes the Sun and stars shine. But the light of quasars is still a mystery. Are they fed by the energy released when stars tumble irreversibly into black holes?

> *Air sank down to make gentle breezes. It extends into the emptiness of space. Set beneath the stars it nourishes their flames.*

What is the origin of the atmosphere? How did Mars and Venus get their atmospheres, which are so different, and which have such different consequences for life? The Earth sits comfortably between the Martian ice age and the stifling Venusian greenhouse.

> *The third level is allotted to water. The ceaseless waves have made great expanses of seas. This fluid, when vaporised, is the essence which nourishes the air. The Earth, on account of its weight, became round and it sank below the other elements. At first it was only a mass of slime mixed with sand, which the water left behind in order to reach a higher level.*

This describes the organisation of matter in the tremendous reaches of space, and the formation of the Earth, precious for our survival At this point we shall insert the amazing advances made in understanding planetary origins by space exploration during the last two decades.

> *As the fluid became more rarefied and dispersed through the air, so the Earth became drier and contracted, forcing the waters to pour into valleys. Mountains came out of the depths of the sea, the land was born in the bosom of the waves, surrounded on every side by the vast ocean.*

Here we put in the fact that as the core of the primitive Earth melts, outgassing shrouds the Earth with its first atmosphere. Furthermore, the first continental plates solidify.

Without those our evolution would have perhaps ceased with the first multicellular marine organisms.

> *The Earth is fixed, because the firmament is separated from it in all senses with an equal distance. The Earth has fallen to its present place and it cannot fall any further. It is at once the central and lowest part of the universe.*

That is no longer valid, but is there not an analogous problem with the entire universe of galaxies? All the galaxies are separating from each other because the universe is in a state of expansion.

Manilius, thank you for this text! It introduced us to a subject that interested you passionately. You have enabled us to explore numerous questions which we can use to tackle the secrets of the universe, which supports us only as a frail but very fine pattern of hoar frost! If we can gain a better understanding of our fate, can we perhaps prevent an unwelcome breath melting away our fragile substance?

Since the seventeenth century, when the astronomer Giovanni Battista Riccioli gave the lunar craters the names we use today, Manilius has had a crater 30 miles in diameter (40 km) in the Mare Vaporum. A well-deserved honour. On 27 May 1985 I climbed to the top of a tall tower in the Sorbonne, capped by a bronze cupola. All around spread greater Paris, a rich carpet of the one of the most extraordinary cultural heritages produced by mankind. Manilius was meeting me that evening. With the help of a telescope I gazed on the immense circular ring which, looking so tiny up there, would effortlessly enclose the priceless jewel of civilisation all around me.

3

A first look at the universe

Powers of ten notation

If you are familiar with powers of ten or exponential notation for very large or very small numbers you can skip this section.

In everyday life we have hardly any occasion to use really big numbers. For quite a lot of transactions the ten fingers are enough. We speak of having one house, two cars, three children, of buying ten apples, or of earning some thousands of pounds or dollars a year. All of these numbers can be written with a few digits. On the other hand, financial newspapers or a finance Minister deal in millions and billions of pounds or dollars.

Science extends far beyond the everyday domain of our senses, and doesn't shrink from large numbers. There is no practical word, at once precise and in general use, to describe the number of synapses in the brain, the number of stars in our galaxy or the number of molecules in a litre of water. To express such quantities we use powers of ten. They have their origin in mathematics, but you don't need to know too much in order to understand their principle.

Quite simply: ten to the power of one is ten or 10; the second power is ten multiplied by itself twice, that is a hundred or 100. The power of three, ten times itself three times, gives one thousand, 1000. Ten to the power of four is ten multiplied by itself four times, yielding 10,000 — the digit 1 followed by 4 zeroes.

Generalising: ten to the power n is 10 multiplied by itself n times, or the figure 1 followed by n zeroes. It is a very useful expression, and it is written 10^n and termed 'ten to the power n '.

Here's an example: in lots of children's playground games you can hear them shout an enormous number: 'one hundred thousand million billion', which is 1 followed by $2 + 3 + 6 + 9 = 20$ zeroes, or ten to the power 20 (10^{20}). It is a colossal number, equal to the number of molecules in a thimbleful of air.

Powers of ten mean we can express big numbers easily. But what about small ones? Instead of multiplying by 10 we divide, and call those 'ten to the power of minus n '. Ten to the power of minus one, that is, 1 divided by ten is one-tenth or 0.1. Ten to the power of minus two is 1 divided by ten twice, that is one-hundredth or 0.01. Ten to the power of minus three is 1 divided by ten three times, one-thousandth or 0.001, etc.

Ten to the power of minus n , that is 1 divided by 10 n times, is written 10^{-n} . So one hundredth thousandth millionth billionth is $10^{-2} \times 10^{-3} \times 10^{-6} \times 10^{-9} = 10^{-20}$. This really is a very tiny number: 10^{-20} metres is one hundred thousand times less than the diameter of the nucleus of a hydrogen atom.

This method of counting is really very powerful. If powers greater than about 20 are used, the results are spectacular. The secret lies in the fact that successive multiplication by ten has a cumulative effect totally different to successive addition. To see this, picture a set of those wooden Russian dolls that nest one inside another, but imagine that each doll in the set is ten times larger than the previous one. Starting with a doll 1 cm tall, the next is 10 cm, and the third is 100 cm (1 metre) while it is 10 metres for the fourth. For the fifth — we drop the idea of making it! And yet five dolls each.of 1 cm use up hardly any wood. The physicist Albert Messiah gave me a very neat example for $n = 6$: when a conference organiser asked him how long he wanted to speak he said 'a microcentury' (10^{-6} century). Adding on seven minutes for a discussion afterwards, that more or less filled an hour in the programme.

There is a further point to note, and I give a illustration of it. Suppose a court of law handed down a sentence of 10^8 seconds in prison and the defence appealed for a reduction to 10^7 seconds; perhaps the court would not trouble to discuss for very long such a trifle, since the difference between 7 and 8 is not much. But 10^7 seconds is four months whereas 10^8 seconds is three years, quite an enormous difference for the prisoner!

Essentially, between 10^7 and 10^8 there is a ratio of ten times. This is the crucial point to grasp in powers of ten notation: a difference of 1 between the values of *n* is ten times, regardless of the value of *n*. There is the same ratio between 10^{20} and 10^{21} as between 10^2 and 10^3, a factor of ten in each case. When we describe the properties of the universe using this notation, keep such ratios in mind.

The size of the universe, in space and time

The universe in powers of ten

Armed with this method of counting, why not use it to explore the universe? In order to do that, imagine a series of snapshots each of which depicts an object ten times larger than the previous one. The first shows a finger, actual size, the second a baby, the third a truck, the fourth a swimming pool, and so on.

We can make a table of objects and their size in centimetres expressed as powers of ten:

10^1 cm	finger
10^2 cm	baby
10^3 cm	truck
10^4 cm	swimming pool
10^5 cm	main street (e.g. Fifth Avenue)
10^6 cm	big city (Chicago, Melbourne)
10^7 cm	large island (Tasmania, Hawaii)
10^8 cm	major country (Mexico, France)
10^9 cm	Earth

10^{10} cm	Jupiter
10^{11} cm	Sun
10^{12} cm	Capella (a star)
10^{13} cm	the Earth's orbit
10^{14} cm	Pluto's orbit
...	
10^{19} cm	distance to the nearest stars
...	
10^{23} cm	size of our Galaxy
10^{24} cm	distance to the Andromeda galaxy
...	
10^{27} cm	furthest galaxies detectable
10^{28} cm	furthest quasars recorded

By 10^{29} cm the progression of distance within the universe is stopped by the cosmological horizon. This is an impenetrable barrier, though not of a material sort, which limits the observable universe, quite independently of any future advances in technology. It has its roots in Einstein's theory of relativity, and is tied in with the expansion of the universe and the impossibility of exceeding the speed of light.

This series of snapshots gives an impression of the immensity of the cosmos. But we haven't finished yet! We will now make a journey in the opposite sense and look at objects progressively ten times smaller than the previous one. Recall the picture with the finger. As we go smaller, the first image is a fingernail, the second the skin, the third a pore, and so on. We construct a new table:

10^{0} cm	nail (10^{0} is of course the number 1)
10^{-1} cm	skin
10^{-2} cm	pore
10^{-3} cm	cell
10^{-4} cm	bacteria
10^{-5} cm	virus
...	
10^{-8} cm	atom
...	
10^{-12} cm	atomic nucleus

10^{-13} cm	proton
...	
10^{-15} cm	intermediate boson
...	
10^{-43} cm	the Planck length.

Our present knowledge of physics will not let us go smaller than the Planck length. There, as with the cosmological horizon, we encounter in the microworld a fundamental and unbreachable barrier.

This set of pictures reveals the fantastic size of the universe. The cosmological horizon is 10^{20} times bigger than the size of the Earth, in other words a hundred thousand million billion times as large, to recall the playground expression.

Equally staggering is the void within that enormous volume. The Earth, the Sun, the planets and the stars only occupy the merest fraction of the available space. The available places are hardly filled, as if a gigantic mess reigns supreme.

This surprise, which we stumble across in the first steps of our scientific exploration, begs the question: why is such a ginormous universe so empty? That is a good question, and straightaway it takes us to the point that our own fate is fundamentally related to that of the cosmos at large. In the future the universe could take two different tracks: on the one hand the expansion of the universe, which commenced in the Big Bang, could continue indefinitely, or on the other hand it could be reversed and the universe could collapse. Present observations are not good enough to discriminate between these two outcomes. We must bear in mind that the formation of the Sun and the Earth and the appearance of intelligent life have taken billions of years.

Even so, if the universe were much less empty than it in fact is, it would be populated with huge numbers of stars and planets. In that case the mutual gravitational attraction would already have overwhelmed the expansion. The universe would have contracted, and *homo sapiens* would not have appeared in time! As it is, it is almost empty and has been in existence for billions of years; mankind now exists and the universe has had the time to become extremely large.

In the first few pages of this book, the tremendous importance that the destiny of the universe has for us has emerged sharply. That importance brings into play, in a very direct way, our own existence and the conditions necessary for our existence. With our knowledge of the distances to the planets and the stars anyone can say casually that the Earth is no more than a grain of sand in the universe. But for me the stakes are higher — more fundamental and more radical.

The time scale of the universe

Now that we have had a look at the distance scale of the universe, let us explore the time axis. Manilius painted a vast canvas in space and time, more than two thousand years ago. Sadly, western philosophers had fixed the creation at 4004 BC, until one hundred years ago.

The genius of Darwin and some of his contemp-oraries brought new vision which showed that one part of the universe does evolve. They claimed that in the biological and geological domains evolution occurs and that it requires vast time scales, far exceeding four millennia.

To explore the time dimension we will start by recalling some of the major events of the last few thousand years:

-300 years	foundation of observatories in Paris and London
-400	Copernicus, Galileo
-2000	Eratosthenes measures the size of the Earth
-4000	invention of writing.

That takes us right through recorded history. Now we move into units of millions of years, in steps a thousand times larger than the table above.

-1 million years *Homo erectus* masters the use of fire
-2.5   *Homo habilis* uses tools
-3.5   the dawn of *Australopithecus*.

That wraps up the prehistoric period. To continue on our journey in time we need to multiply the basic unit again by one thousand, so now we are dealing in billions of years.

-0.5 billion years explosion of life in the Cambrian era
-1.5   first eucaryotic organisms
-3.5   first bacteria
-4.5   formation of the Sun and the Earth
-8   age of the most distant galaxies
-15   the Big Bang.

The boundary of history is reached with the geological eras, followed by the cosmological era. This way of compartmentalising time into units each of which is 1000 times the size of its predecessor is impressively compact. The age of the universe is much longer, by a factor of 250 000 times, than the pre-Darwinian estimate. Physicists prefer to measure time in seconds, and expressed that way the age of the universe is close to 10^{18} seconds.

Physicists can study phenomena of very short time duration. The results obtained by high-speed photography are well known, such as the pictures of a rifle bullet in flight. In terms of short duration, that is nothing compared with the transient phenomena now detectable in laboratories. For example, flashes of ultraviolet light as brief as 10^{-14} seconds hold the current record. One such flash, travelling at the speed of light, is itself less than a millionth of a centimetre wide, less than the size of a bacterium. It is so brief that it represents only eight wavelengths.

In conventional physics, shorter times than this are encountered when we discuss the frequency of vibration of electromagnetic waves. For visible light the wave frequency is equivalent to one vibration in 10^{-15} seconds. But the shortest wavelengths, like those of X-rays and gamma rays, have vibration times of 10^{-22} seconds. In nuclear physics we encounter even shorter intervals, typically the time for light to cross the atomic nucleus, which is 10^{-23} seconds. In the domain of certain elementary particles, such as the bosons, the scale descends to 10^{-26} seconds.

finally, within the last ten years, physicists engaged in the quest for the ultimate structure of matter have developed theories such as Grand Unification with infinitesimal time intervals such as 10^{-35} seconds. The detailed experimental verification of such theories would need particle accelerators like those in CERN, Geneva, but of an enormous size: not tens of miles or kilometres, but dimensions approaching those to the nearest stars. Needless to say, such giants will never be constructed, certainly not by the present generation!

How can we test the theories then? Where can we find the physical conditions, or the enormous energies, that we need in order to verify particle physics? This brings us to another miracle of scientific discovery in the last few years: the conditions existed in the first few moments of the Big Bang. Right after the Big Bang the particles filling the universe had an enormous amount of energy. This has led to a fruitful alliance between physicists who study the ultimate structure of matter and astronomers who want to penetrate the mysteries of the first few seconds of the universe's existence. One such collaboration has greatly benefitted both disciplines by shedding new light on the structure of matter and on the behaviour of the universe, and taking us closer to the secrets of its fate. This close meshing of different branches of physics is at 10^{-35} seconds after the onset of the Big Bang. That might seem a tiny interval of time to us, but it determined the future of the universe, and all of its evolution.

Is it likely that physics will, in the future, let us get even closer to the origin? That is an open question, but there may be an impenetrable barrier set by the Planck time. This is the travel time for light across the Planck length, and its value is 10^{-43} seconds; the question is of great scientific and philosophical interest, and is actively researched.

It seems remarkable that we have come across this confrontation of infinitesimal elementary particles and the largest structure of them all, the universe, and that at the instant of the Big Bang the immensity of the cosmos is unimaginably small. But the strangest thing of all is that the briefest of brief moments at the start of the universe was of major importance for its destiny and its observed state 15 billion years later. Here is just one example: matter itself, so vital for us, is present in just the right quantity, not too much and not too little, and that outcome was determined in the Big Bang.

Building blocks of the universe

The structure of the universe

Imagine a person trying to find out about their own country simply by looking at things on a microscopic or a macroscopic scale. Let's speculate on what a very short-sighted parachutist might find, dropped randomly a dozen times. He would probably conclude that there was little more than blades of grass, unless by chance he saw a flower or an insect as well.

If, instead of being content with this very local view of the 'universe' he wore eyeglasses, he would examine a much wider sample at each drop. He would find other components: trees, stones, and birds perhaps. And if he stood up he would scan a vast panorama of houses, hills and tall buildings maybe. This would be the result of the 'microscopic' examination.

Suppose he chooses the other 'macroscopic' method and stays in the plane in order to see everything at one pass. In an agricultural country he would see vast expanses of green, interrupted by darker patches criss-crossed with tracks and furrows.

For people who live at ground level, space is essentially two-dimensional: it has length and breadth. So far as they are concerned, an aviator exists in another dimension, specific to him and inaccessible to them; this is the third, or height, dimension. (Of course, people used to the huge apartment and office buildings in large cities are quite accustomed to expressing themselves in three dimensions, but our story concerns imaginary beings who do not have that experience.) For such people, the parachutist would seem like a phantom who suddenly materialises out of nothing onto their space (the ground) of two dimensions.

This rather contrived story is an analogy to help with the exploration of the universe. The space containing the universe has the three dimensions of our everyday experience: length, breadth, and height. The surprise parachutist would, in our terms, come from a fourth dimension. If he were very myopic what would he see? Nothing! Absolutely nothing! The main content of the universe is in fact empty space. The commonest objects in space are photons, or quanta of electromagnetic radiation, from the famous cosmological microwave background radiation at 3°K, which is a fossil of the Big Bang. They are quite sparse: about 400 photons per cubic centimetre.

If, by a sheer fluke, the parachutist jumped out in a galaxy (one chance in a thousand), and if within that galaxy he had the incredible luck to be very near a star (one chance in a thousand billion), he would find another population of photons, this time from the nearby star.

Continuing this lonely quest, he would eventually encounter atoms of hydrogen, but only within galaxies, because intergalactic space is almost a perfect vacuum. He would also find interstellar gas molecules, and grains of interstellar dust about one micron in diameter with an average density of only one grain in a cube of side 100 metres. Lastly, with a chance of one in 10^{24} he finds he has popped up inside a star, surrounded by matter at a temperature of some millions of degrees. What about planets: could he find any of those? If we assume that all stars have planets like the Sun he has only one chance in 10^{27} of landing on one, which is about the same chance as finding a needle in a cubic haystack 1000 kilometres in length!

All of this is pretty frightening, but what if we ask the chance of our hypothetical parachutist materialising on Earth from a random jump into such a vast and tenuous universe?

The answer is one chance in 10^{57}, which is equivalent to finding a bacterium in a volume with the dimensions of the solar system. So, our imaginary parachutist, short-sighted remember, has essentially no chance of ever finding the Earth.

Thus far, his myopia has not stopped him from locating 'microscopic' objects: photons, atoms, and molecules, as well as planets and stars. But for 'macroscopic' objects he certainly needs eyeglasses. What will he find then? Dropped randomly in space, he will end up in the vast empty reaches of intergalactic space, and will be able to see distant galaxies, the nearest being at a distance of a million light years or so. These fluffy objects, shining wanly, will look to the unaided eye like the Andromeda galaxy does to us, if you can imagine it by itself without the bright starry sky.

If the parachutist finishes up in a galaxy, the view will be very much like our own starlit nights: myriads of stars, star clusters and gaseous nebulas. But he probably would not see any planets. That is because they are invisible at distances of more than a few light days from a star. The tiny chance of encountering the Earth arises because the parachutist has to land in our Galaxy, out of all the tens of billions of other galaxies in the immensity of space, and then select our Sun from hundreds of billions of stars.

Suppose instead we explore the cosmos like the parachutist, but we stay inside the airplane. Imagine surveying the universe from a hypothetical exterior, as if we are looking on it from a fourth dimension. This is analogous to the third dimension that would enable one to see a huge expanse of a two-dimensional country at one sweep. From the imaginary fourth dimension, a vast and majestic three-dimensional panorama extends as far as the eye can see: billions of galaxies grouped into clusters, and the clusters arranged in superclusters. On a scale of one billion light years, the superclusters themselves trace a delicate filamentary structure, a lacy vista of areas richly populated with galaxies, separated by immense voids.

At around ten billion light years we encounter the limits of observation, imposed by the cosmological horizon.

However, an observer in the fourth dimension would have a less constrained view of things. He could find out whether, beyond the horizon, the cosmos continues, ten or even a thousand times further and see if space is curved and closes back on itself, like the surface of a sphere, only it would be the three-dimensional surface of a four-dimensional sphere. In the limit of our flight of fancy, he might go into universes disconnected from our own, perhaps very different, and perhaps the majority of them would not have any intelligences capable of asking about the cosmos and their place within it...

Although there is no such person as this imaginary observer, theorists on Earth have tried to replace him with established theories, supported by conventional physics and observation, such as Einstein's General Theory of Relativity, or more recent and speculative notions which are now taken seriously, such as Grand Unification or the Inflationary Universe. Such theorists are bringing exciting perspectives to cosmology.

Matter in the universe

One of the oldest established technologies of mankind is the making of stone tools. Stretching back for some two or three million years this is much older than the equally important advance when fire was mastered, perhaps one and a half million years ago. This long-established use of stone goes back further than our own species *Homo sapiens* to *Homo erectus* and before that *Homo habilis*, who was the presumed inventor. For hundreds of thousands of years this technology stayed on a plateau without any significant evolution. For example, the manufacture of flint axes was repeated over tens of thousands of generations without the slightest technological advance, as father passed the skill to son, son to grandson..., always the same skill and technique. It is not all that surprising that the idea of the hardness of flint, and by extension, the durability, solidity, density and impenetrability of all solid matter should be so deeply embedded in our our minds and our culture. In the modern world it takes a real effort, even after years of study, to appreciate that matter, even stuff as hard as flint or diamond, is nearly empty!

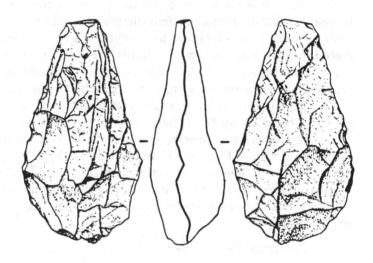

figure 1. A palaeolithic stone axe

As a matter of fact, as theory and modern investigations of the nature of matter progress, so they seem to get further from any tangible reality. Our situation is somewhat like that of a sailor bowling confidently along in the ocean breezes who suddenly sees a manufacturer's warning saying that the craft is made of light and fragile material that dissolves in water!

The nature, structure and stability of matter, as well as its fate, are of major importance. Furthermore, the everyday mechanics of the universe — the creation of the chemical elements and the release of energy in stellar interiors, so vital for our existence — is strongly dependent on the properties of matter. To study the story and destiny of either the universe or the matter within it, we need to study both.

The extraordinary variety of substances that we call 'matter' are the result of varied and innumerable combinations of just 92 different atoms. It is just one sign of the relentless progress of science that we can explain this vast diversity by means of such a small number of elements. The number of ways of combining 92 elements in principle is mind-boggling: if we combine them every which way there exist 10^{142} variations!

These 92 elements, from the lightest hydrogen to the heaviest uranium, form the molecules. Atoms are very tiny, about 10^{-8} centimetres in diameter, or one ångström in the everyday units of atomic physics. Two hydrogen atoms join with one oxygen to form water; three hydrogen and one nitrogen give ammonia; and four hydrogen atoms linked to one carbon atom make a molecule of methane.

The molecules can be found in various states, of course: water molecules can be found as ice, liquid water, or water vapour, depending on whether the individual molecules are frozen into a crystal lattice, are able to move freely and randomly, or are separated far from each other. Which of these states the molecules are in is a function of the ambient temperature and pressure. Under atmospheric pressure, water is ice below 0°C, is in the form of liquid from 0° to 100°C, and is vaporised above 100°C. But at much lower pressure, say one-tenth of an atmosphere as exists on the planet Mars, only ice or vapour can exist and there is no liquid state. Ammonia and methane behave similarly, although the values of temperature and pressure for a change of state are different to water. The solid or icy forms of water, ammonia and methane play an important role in the planets and satellites of the solar system, as well as interstellar dust. They are crucial for the formation of complex organic molecules.

Organic molecules are important for our fate. Their name 'organic' derives from the fact that they are found principally in living organisms. More properly we should call them molecules based on the chemistry of carbon. The impressive variety of these molecules stems entirely from specific properties of carbon atoms, which are able to link into enormous frameworks and structures. Molecules with hundreds and even hundreds of thousands of carbon atoms can be assenbled, along with huge numbers of hydrogen, oxygen and nitrogen atoms. These structures can be grouped into chains like the hydrocarbons methane, ethane, propane ... with 1, 2, 3 ... carbons and 4, 6, 8 ... hydrogens; or into rings like benzene, diphenyl, ...; rings and chains; huge and complex chains; right up to the double helix of DNA, the carrier of the genetic code.

The universe has life as we know it because of the extraordinary diversity of carbon chemistry. It makes us feel rather insignificant to know that the cosmos could have had a completely different outcome. Where have the 92 elements come from? This is a sufficiently large number for us to hope that the 92 different atoms could be assembled from a smaller number of more fundamental building blocks. Manilius did well with the four basic elements of the classical world: fire, air, earth and water. In fact, all atoms consist of electrons surrounding a central nucleus that is made of protons and neutrons. There are just three components: electrons, protons and neutrons.

The complexity of matter arises from this simple state of affairs. These three components of the atom bring us straightaway to very small dimensions. Protons and neutrons, which are collectively called nucleons, are of order 10^{-13} centimetres in diameter. This means they are 100 000 times smaller than atoms. Jammed close together, they form the nucleus. What differentiates the 92 natural elements is quite simply the number of protons in their respective nuclei. This number is 1 for hydrogen and 92 for uranium; for carbon, nitrogen and oxygen the atomic number (number of protons) is 6, 7 and 8, respectively.

The origin of this matter is bound up in the origin of the universe itself. Only minutes after the Big Bang most of the matter in the universe consisted of electrons, protons, and the nuclei of helium atoms. Processing in nuclear furnaces inside exploding stars has created all natural elements heavier than hydrogen and helium.

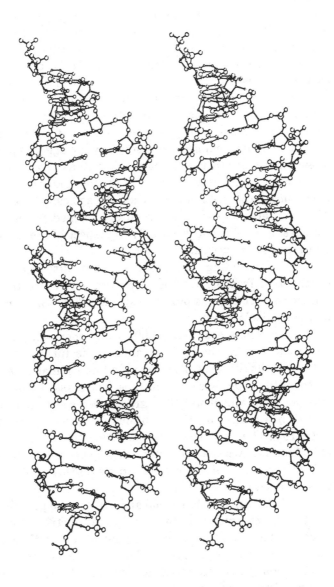

figure 2. Stereo view of part of the DNA helix.

In addition to the protons, atomic nuclei also contain neutrons. The number of neutrons in naturally occurring nuclei is equal to or somewhat greater than the number of protons. For example, carbon has six protons and usually six neutrons; however, one per cent of natural carbon has seven neutrons in the nucleus. Can more neutrons be found? In exceptional cases yes: carbon atoms bombarded by high-energy cosmic rays acquire an extra neutron to make carbon-14. This is unstable and the atoms decay with a half-life of about 5000 years. This has led to the development of a very useful technique for determining more precisely the age of objects just a few hundred or thousand years old, such as ceramics, which had carbon-14 atoms at the time of their formation. By seeing how much carbon-14 remains today, the age can be calculated. These different species of carbon are termed isotopes. The isotopes of an element differ only in respect of the number of neutrons in the nucleus.

Surrounding the nucleus, at a distance of 10^{-8} centimetres from it, are the electrons. There are the same number of electrons as protons, between 1 and 92, depending on the particular element.

Nucleons are nearly 2000 times more massive than electrons, and so the mass of an atom is furnished almost entirely by its nucleons. The mass of a lump of matter is that of its constituent protons and neutrons. The mass of a nucleon is extremely small: about 10^{-24} grams. We see that ordinary matter has a huge number of atoms per cubic centimetre, of order 10^{23}. This number is consistent with the atomic diameter of 10^{-8} centimetres since 10^{23} atoms closely packed, as in a liquid will fill a volume of one cubic centimetre; this volume has a mass of one gram for water.

In addition to dimension and mass, these particles have electrical properties: protons carry an electric charge +1, neutrons have charge 0 (they are neutral, hence their name), and electrons have charge -1. These are arbitrary units of electric charge. As an entity, an atom in its normal state is electrically neutral, since the numbers of protons and electrons are equal and they carry opposite charges.

Does this notion of electrical charge have reality? With this concept, we can compute important interactions that occur in nature, the electromagnetic interactions. In the same way as the concept of mass lets us calculate gravitational interactions or analyse the dynamic behaviour of particles in a gravitational field, so electric charge lets us follow the behaviour of charged particles under electro-magnetic forces.

The chemical properties of atoms are determined by the retinue of electrons. Those furthest from the nucleus are the most important, because they are most readily influenced by external circumstances. Through the medium of electromagnetic forces between atomic electrons, two hydrogen atoms can combine with one oxygen atom to make a stable structure, one molecule of water. In exactly the same way atoms of oxygen and silicon (number 14 in the periodic table of the elements) combine in a rigid and resilient crystal lattice to give flint its great strength, discovered long ago. Today we have a completely satisfactory picture of the atomic theory of matter.

The glue of the universe

Now that we know what matter is made of we have to answer the question: how do its constituents interact.?We have found the individual bricks but where is the cement to hold everything together? There are four interactions known to affect matter. They are the gravitational interaction, the electromagnetic interaction, and the strong and weak nuclear interactions.

Gravitation

Of these four interactions, gravitation has been recognised the longest. It was discovered by Newton at the end of the seventeenth century. He formulated a law to describe it: two masses attract each other in direct proportion to the product of their masses and inversely as the square of the distance between them. Think of two equal masses, say one kilogram each, separated by one metre. The mutual force between them can be denoted by a universal constant, G. In general, for two masses m_1 and m_2 kilograms separated by r metres, the force F between them is given by the constant G, multiplied by m_1 and m_2 successively, and then divided twice by r. We can express this as a formula:

$$F = Gm_1m_2/r^2.$$

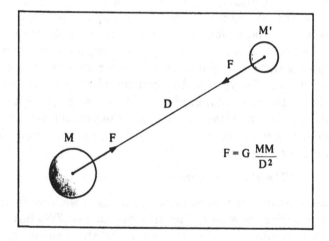

figure 3. Newtonian interaction between two point masses

G is a universal constant which determines the strength of the gravitational force. Its value is such that a human with a mass of 70 kilograms situated 6387 km from the centre of the Earth (that is to say, on the surface), which has a mass of 6×10^{21} kg, experiences a force which is equal to the weight of 70 kg. On the Moon, taking into account the smaller mass and radius, the human would be pulled with a force which Newton's law gives as 12 kg. He would have 12 kg of weight, but 70 kg of matter. The value of G is a constant of nature, whose value is determined by experiment. Theorists have pondered the question as to whether the value is immutably fixed, by suggesting that in the remote past history of the universe G could have been different.

Newton's law of gravitation has been extremely fruitful. By applying it to the solar system, Newton easily explained the three empirical laws of planetary motion, discovered 70 years earlier by Johannes Kepler, who analysed the observations of Tycho Brahe. Its success led to the establishment of a branch of applied mathematics known as celestial mechanics. A good account could be given of the structure of stars, the shapes of star clusters, galaxies, and clusters of galaxies. The Newtonian theory did not meet any major difficulty until the early twentieth century: it proved impossible to give a complete explanation of the observed precession of the major axis of Mercury's orbit round the Sun. Einstein's theory of General Relativity gave the full solution. Nevertheless, Newtonian theory is good enough for a first approach to the cosmos.

The electromagnetic interaction

The electromagnetic interaction, which was basically discovered in the nineteenth century, makes itself felt principally through electric and magnetic forces. The simplest example of this is the electrostatic force, which follows a law very similar to Newtonian gravitation. Two electric charges experience a mutual force in direct proportion to the product of their charges and inversely proportional to the square of the distance between them. There is one important difference: this force is one of attraction, as in gravitation, if the two charges are of opposite signs. If they are of the same sign then the force is one of repulsion. This important difference is encountered in a variety of circumstances. It enables us to explain how negatively charged electrons are attracted by the protons in the nucleus, and thus orbit around it, somewhat like the planets around the Sun.

Magnetic effects are another major difference. If an electric charge moves in a circle (like an electric current in a coil of wire) it creates a magnetic force (and the coil of wire behaves like a magnet). A magnetic force alters the trajectory of a moving electric charge; by this means, bunches of charged particles are constrained to move in circles in particle accelerators. Magnetic fields also act on other magnets; electric currents in the Earth's interior affect magnetic compasses. If electric charges are trapped and made repeatedly to change their direction of motion, the fields they create change their sense also, and this gives rise to electromagnetic waves which travel at the speed of light. Currents oscillating thousands or millions of times a second in coils generate radio waves from one kilometre to one metre in wavelength.

In the nineteenth century James Clerk Maxwell unified electrical and magnetic phenomena into a single electromagnetic theory. This shows that the interactions and forces depend on position as well as motion. Just like Newtonian gravitation, Maxwell's electromagnetic theory is good enough for a first tour of the cosmos.

The strong nuclear interaction

Now we have a conundrum: how can the protons of the nucleus, all of which have positive charge, stick together in spite of their electrical repulsion? It is true that they experience mutual gravitational attraction, but the value of the constant G is so small compared to electrical forces that the electric repulsion overwhelms it by a factor of 10^{36} times. For the nucleus to hold together there has to be a mighty force, capable of defeating the electrical repulsion. This is the nuclear force, or strong interaction, discovered in the 1930s.

The name strong interaction arises from the intensity or strength of the interaction, but its major characteristic is its very short range, in marked contrast to gravitational or the electromagnetic interactions, which operate at great distances althought their effect is diluted by the inverse square of the distance. The range of the nuclear interaction extends no more than 10^{-13} cm and is practically constant over that domain. The nucleons behave like hard spheres: beyond 10^{-13} cm they do not interact because to do so they need to be almost in contact. A characteristic of this behaviour is that a given nucleon will interact only with its nearest neighbours in the nucleus, at half a dozen other nucleons. Similarly, in a bag of marbles, any one marble touches only a handful of nearest neighbours.

There is another consequence of the short range of this interaction. The energy which a nucleon possesses by virtue of the fact that it is in a nucleus is not a function of the size of the nucleus, since only a handful of other nucleons can affect it, regardless of whether the nucleus is big or small. These properties mean that the matter within an atomic nucleus behaves in many respects like a liquid of fantastic density. That density is the mass of a nucleon divided by its volume: 10^{15} grams per cubic centimetre, or 10^{15} times the density of water.

This density is staggering even for things as small as atomic nuclei. Can we imagine the discovery of macroscopic objects with densities of billions of tonnes per thimbleful? Yes. Predicted in 1934, these objects are the legendary neutron stars and pulsars, collapsed stars in which a ball 10 km across has the entire mass of the Sun (which is 1.5 million km or nearly 1 million miles in diameter).

The weak interaction

The nuclei of some atoms are unstable; quite spontaneously they emit radiation. We call this radioactivity. For example, alpha decay involves the emission of an alpha particle, which consists of two protons and two neutrons and is identical to the nucleus of a helium atom. In such a radioactive transition the final state of a new atomic nucleus and an alpha particle is more stable than the initial state. This extra stability has come about because the nucleus has re-arranged itself and got rid of the alpha particle in the process. This sort of radioactivity is accounted for in the physics of the strong interaction.

Beta radioactivity or beta decay cannot be explained that way. When beta decay was discovered right at the start of the twentieth century the particles responsible were shown to be electrons. What happens is that a neutron in the nucleus transforms to a proton, by expelling negative electric charge as an electron. There is no change in the number of nucleons, but the atomic number and the charge increase by one. About fifty years ago a variation was found: the particle emitted is not an electron, but rather a positron. This is the anti-particle of the electron; it has the same mass but carries a positive charge. In such a case the nucleus loses electric charge.

An interesting phenomenon is observed in beta decay. The initial state and the final state of the parent nucleus have different energies, but the outgoing electron does not take all of that energy with it, which seems to violate the principle of conservation of energy. The explanation is that there is a further invisible carrier of energy without electric charge and with almost zero mass which is also emitted in radioactive decay. This is the neutrino, first detected in 1956.

Enrico Fermi was responsible for deriving a theory of beta decay on the strength of the above observations. His theory has similarities to the electromagnetic theory but is much more complicated. It is the theory of the weak interaction. Its characteristic features are that it is a very weak and very short range one, one hundred times shorter than the strong interaction. For example, via the weak interaction a neutrino passing right through the Earth has only one chance in ten thousand of interacting with a nucleus.

An example of the weak interaction at work is given by the decay of a neutron into a proton. Free neutrons (those not part of an atomic nucleus) behave like nuclei subject to beta decay. With a half life of just 20 minutes they change their state; a neutron becomes a proton by emitting a neutrino and an electron. In some very special circumstances the opposite reaction can take place: in the intensely hot interiors of certain types of stars protons and electrons combine to form neutrons, and they emit neutrinos in the process. Under their mutual gravitational attraction the neutrons form an incredibly dense neutron star. The escaping neutrinos transfer their energy to the outer layers of the star which are hurled into space in a supernova explosion.

Not all particles can participate in the weak interaction. By definition only those sensitised by their possession of a 'weak charge', analogous to the electric charge of particles subject to electromagnetic forces. The neutrino only responds to the weak interaction because it has no electric charge, so it cannot respond to the far stronger electromagnetic interaction.

Galaxies and stars

Galaxies

In many science museums and planetaria you can see huge murals and models of galaxies. One of my favourites is in a museum in Paris where they have a fluorescent model of a spiral galaxy suspended by invisible threads in a dark diorama. When I look at it I get a creepy feeling: it brings home to me the mystery and majesty of the largest objects in our universe. On the scale of such models, you the spectator have a height of one million light years. What is particularly striking is the softness of the light, almost translucent, that seems to come from everywhere and nowhere. In fact this diaphanous glow is the combined light of tens or hundreds of billions of stars, seemingly merged into one mass when viewed from so far, but in fact separated from each other by much greater distances than their individual diameters.

We can imagine the Sun, its planets, and ourselves on the Earth, being members of a spiral arm towards the edge of the galaxy. From such a vantage point we see the universe of galaxies, and we look edgewise through our own Galaxy, which appears to us as a misty veil across the heavens, the Milky Way. To gaze on that ancient light is to see the glimmer of a whole galaxy. The solid Earth cuts out half the view, but with imagination we can continue the arc above to a full circle beneath our feet. If you do just that on a crystal clear night, with the sky as black as velvet so the stars seem near enough to touch, you can get the profound impression that you are living in a gigantic galaxy. Such an unforgettable feeling is one of the treasures of the cosmos.

A few degrees away from the Milky Way, on the south side in the constellation Andromeda, there is a faint smudge of light that looks rather like the Milky Way. This is another big galaxy, actually, and it is the nearest to our own. It contains one hundred billion stars, spread through a system 100 000 light years in diameter, situated two million light years away. These bare statistics, so essential for the progress of science and for turning this great vista into something quantitative, do not detract in any way from the drama and beauty of the universe.

Nature is full of wonderful surprises beyond the scope of our own rather limited senses. To look at the detail in galaxies it is better to take photographs with large telescopes. However, these give a hard and contrasty image, which is not quite the same as the real thing. So why don't astronomers peer through their telescopes? The reason is that to see a typical galaxy by eye you need a huge telescope. There are not very many of those, and they are used exclusively with instruments that are much more sophisticated and sensitive than the human eye to gather the data. It is in fact rather difficult to take even a quick peek through a big telescope because the equipment is in the way. When using a telescope myself I have often looked through the eyepiece but only to point the telescope accurately on the object. Generally, when the instrumentation is being pushed to the limits, in order to roll back the frontiers of our knowledge, the galaxies under investigation can be a hundred times further away than Andromeda, and they are barely discernible at the telescope. On one unique occasion I stole a glance at a nearby galaxy through a big telescope, during a ten minute gap between two observations. What I looked at was galaxy M51 in *Canes Venatici* (The Hunting Dogs). This magnificent spiral has a neighbouring irregular galaxy enmeshed in one of its spiral arms. Just as in photographs and models of galaxies, I could pick out a sensation of colour. The nucleus of the galaxy was reddish, like its companion, whereas the spiral arms were blue. I could see extraordinarily fine detail with outstanding clarity. The fragility of it all left an unforgettable impression in my mind.

In terms of the scale of the universe that is now accessible to large telescopes these galaxies are mere grains of dust, uncountable in the tremendous depths of the cosmos. If you let a speck of dust represent one galaxy, the observable universe would be about 10 metres across. Within it would be a few tens of kilograms of dust. Galaxies are the most obvious constituents of the cosmos.

Stars make up the major part of the mass of a galaxy. Galaxies, giants and dwarfs alike, contain between 1 and 100 billion stars, separated from each other by a few light years. The stars within a galaxy are held together by mutual gravitational attraction. Each star is attracted by all the others in accordance with Newton's law. The swarm of stars is very concentrated at the centre, less so at the edge. Rotation of this cloud of stars causes it to become circular and flattened, like a disc, on account of centrifugal forces. Every star follows a circular orbit around the dense central part of the galaxy, which is an ellipsoidal bulge with a very concentrated mass at its centre, the galactic nucleus. Our Sun is 30 000 light years from the centre of the Galaxy and, travelling at 250 kilometres per second, it takes 250 million years to complete one orbit. The discs of spiral galaxies are made particularly brilliant by the presence of very young and extremely energetic stars in the two spiral arms which emerge from the nucleus. Our own Galaxy and the Andromeda Galaxy are typical spiral galaxies.

If the rotation is only very slight, the disc of stars does not form and the central bulge remains. These are the elliptical galaxies, among which are found the most massive galaxies known. The nearest giant ellipticals are in the cluster in Virgo, about 30 million light years away. Other ellipticals are small, or dwarf, for example the two satellite galaxies of the Andromeda Galaxy.

Some galaxies do not have any well-defined shape. These are the irregulars. Perhaps their chaotic structure is a remnant of a particularly turbulent birth. The nearest examples of these are the Magellanic Clouds, two dwarf irregular galaxies, satellites of the Milky Way, which are about 200 000 light years away. They are prominent naked-eye objects in the southern hemisphere and are named for the sixteenth century navigator who first described them.

Radio galaxies

In between the stars, in interstellar space, there is a gas which is three-quarters hydrogen and one-quarter helium. Apart from the effects of its gravitation, this gas is the site of some fundamental interactions. Inside an atom of hydrogen, the electron can interact with the proton as if they were two bar magnets. This interaction leads to the emission of a photon with a characteristic wavelength of 21 centimetres. These radio waves, which are of atomic origin, have enabled us to find out fundamental properties of the interstellar gas: its density, its velocity, and its distribution.

Interstellar atoms in the vicinity of a hot star are bombarded by ultraviolet radiation from the star and this tears the electron away from the atomic nucleus. This process, termed ionisation, results in a mixture of fast-moving free electrons and free protons, in place of the neutral atomic gas. The electrons and protons in rapid motion form a gas at a thermal temperature of 10 000°C. When a high-speed electron has a close shave with a proton, the attractive electric force between the two causes the electron to travel on a curved path, just like electrons in a coil of wire. The laws of electromagnetism show that the electron will emit electromagnetic radiation. Generally, in astronomical situations, the electromagnetic waves appear in the radio region of the spectrum, at centimetre and metre wavelengths. The wavelength at which emission is strongest depends on the velocity of the electrons, which in turn is a function of the gas temperature.

Thermal emission of this type can lead to galaxies being radio sources. But there is a more dramatic mechanism, synchrotron emission, and in galaxies where this takes place extremely violent conditions accelerate electrons to much higher velocities and energies than in the case of thermal emission. Supernova stars, which explode with spectacular violence, produce electrons with speeds close to that of light. Inside galaxies, turbulence or rotation within the ionised gas creates magnetic fields. Although these fields are very weak, only 1/100 000th the strength of the Earth's magnetic field, they extend for vast distances in galaxies and can have far reaching effects.

Energetic electrons spewing out of supernova explosions travel through the weak magnetic environment of the interstellar medium, where their paths are bent and curved. As a result, the electrons send out radio emission at wavelengths from 1 to 100 metres. If the magnetic field is stronger, as it may well be close to the supernova, the wavelength is shorter; it may even be smaller than one micron, which is in the visible part of the electromagnetic spectrum. Exactly this is observed for the supernova remnant in the Crab Nebula. This is the relic of an explosion observed by Chinese and Korean astronomers in AD 1054. Such radiation is also produced in synchrotrons, used by physicists, and indeed it was first observed in that type of particle accelerator.

Atomic, thermal, and synchrotron radiations each cause some galaxies to be detectable with radio telescopes. Among such galaxies we find extremely useful tools for exploring the universe: radio galaxies and quasars. These are found at very great distances, deep in the universe of long ago. By means of radio waves, astronomers can study with relative ease galaxies that are at the practical limit of optical telescopes. There is a famous example of this: towards the end of the 1950s one of the strongest radio galaxies was shown to correspond with one of the faintest galaxies visible with the 200-inch telescope on Mount Palomar. Pushed to the limit, radio wave observations take us to very great depths in the universe, to distances where the weakest radio sources observed cannot be matched to any visible object, the corresponding galaxy being simply too far away.

Nevertheless, the lack of optical data puts astronomers at a disadvantage: the intrinsic properties of radio galaxies are rather poorly known because their distances are hard to estimate. Only the number density of distant radio sources indicates that beyond the visible universe space continues. It does not peter out into a vacuum, nor give way to a hypothetical centre densely populated with galaxies.

Unfortunately, given the present state of our knowledge, particularly about the generation of radio emission in extragalactic objects, we cannot say how far 'beyond' is when measured in billions of light years; we can only guess that it is somewhere between 6 and 10 billion light years beyond the visible range.

Quasars

Happily, a discovery that was not initially expected by radio astronomers enabled optical astronomers to take up the challenge of probing the depths of the universe. To appreciate this discovery, it is worth just reviewing the Doppler effect.

When an object that is emitting electromagnetic radiation travels away from us, the wavelength of the radiation we receive is increased, and it is decreased for an approaching emitter. The greater the velocity relative to the observer, the greater the change. For example, sodium atoms emit yellow light with a wavelength of 0.59 microns (5900 ångströms), and this wavelength is increased by 10% if the atoms travel away from an observer at 10% the speed of light, 30 000 kilometres per second. (Incidentally you can see this sodium light for yourself by throwing a small pinch of table salt into a flame.) At such a speed of recession the emission seems red, not yellow, because the waves have been shifted in wavelength down to the red part of the optical spectrum. In the particular case of a recession velocity of 34 110 kilometres per second the redshifted yellow radiation from sodium coincides exactly with the normal wavelength of one of the principal transitions seen in interstellar hydrogen clouds. This is the famous 'H-alpha' red light of hydrogen at 0.66 microns (6600 ångströms). The Doppler effect arises from laws governing combination of motions; the analogy frequently given is the observed drop in the pitch of a siren as a train or police car rushes by. The redshifts in the spectra of galaxies are interpreted by most astronomers as being due to the Doppler effect, with the implication that redshifted galaxies are travelling away from our Galaxy.

In about 1960 radio astronomers discovered that some of the strong sources of synchrotron radio emission did not correspond with galaxies; instead their positions coincided with starlike objects on photographs (from which they derive the name quasar, a contraction of quasi-stellar). The spectra of these apparent stars showed lines strongly shifted to the red. The interpetation that these redshifts were due to the Doppler effect implied that the recession velocities were of the order of thousands of kilometres per second, speeds far in excess of those known for the stars circling our Galaxy, where the characteristic velocity is hundreds of kilometres per second. If these velocities are plugged into Hubble's law, in which the recession speed is proportional to the distance from us, we come to the conclusion that the quasars are at very considerable distances, billions of light years in fact.

The nearest quasar, 3C 273, recedes at 50 000 kilometres per second and is about 3 billion light years away. Despite that, it is bright enough to be visible in a first class *amateur* telescope. Even if it were 100 times more distant, we could still see it with the world's largest telescopes. These intergalactic beacons have an intrinsic luminosity equal to one hundred giant galaxies. They enable us to probe the furthest recesses of the universe. At the same time, they let us see into the remote past. An astronomical object 10 billion light years from the Earth is viewed in a part of the universe which is 10 billion years younger than our local neighbourhood because the light has taken 10 billion years to reach us. So, quasars not only let us see to great distances, but also look back at the history of the universe.

In fact, there aren't any quasars 100 times further away than 3C 273; the observable universe doesn't go 100 times further! The cosmological horizon, an absolute limit imposed by the expansion of the universe and the theory of relativity, stops observation well short of that distance. There is a further reason why astrophysicists and cosmologists are making special efforts to find out what could have happened in the past: it is quite possible that in the first two billion years of the existence of the universe there were no quasars because they had not had sufficient time to form. The observable quasars take us no nearer than one or two billion light years from the cosmological horizon. The fact that quasars are not permanent features of the cosmos is indicated by studying the nearest quasars. They are much less common, in terms of numbers per unit volume of space (after correcting for the expansion of the universe), than the quasars a few billions of light years away, perhaps by a factor of 100. They seem to be objects that took a certain amount of time to form, and then have a rather brief middle age of just a billion years or so. They are a species of the cosmic zoo that appeared a little after the Big Bang, and they are now becoming extinct, as the universe outlives them.

Nevertheless, because quasars are such strong optical and radio emitters, they enable our present telescopes and detectors to probe space up to 80% of the distance to the cosmological horizon and to explore past history to within two billion years after the Big Bang. Like lightbuoys on a vast spacetime ocean they send us signals about poorly charted regions. And for that reason astronomers have taken an intense interest in deciphering their messages in the 25 years since they were discovered.

Unfortunately, there are a lot of problems and the situation is similar to that for powerful radio galaxies. In spite of the optical work, our knowledge of quasars is not enough to enable us to work out the physical processes taking place in these objects. These beacons of uncertain origin, sending out signals in a mysterious way, are almost useless for charting deep space.

A major puzzle still is the source of the prodigious energy that they emit, not only in the radio region of the spectrum, but also light, ultraviolet rays, X-rays and even gamma radiation. Data acquired recently, using sensitive detectors on large optical telescopes such as those at Palomar and on Hawaii, or giant radio telescopes like the Very Large Array in New Mexico indicate that the 'black box' generating this fabulous energy is extremely compact and is located in the nucleus of the galaxy. There are a couple of fashionable hypotheses concerning the nature of this black box. Perhaps there is a cluster of stars that is very tightly packed at the centre of the galaxy in which the stars are destabilising each other, or even colliding, so that an on-going series of supernova explosions is triggered. The other idea is that there might be a black hole with a mass several tens of millions of times the mass of the Sun. Such a black hole, situated at the centre of a galaxy, devours every star that comes too close: they literally disappear irreversibly into the hole. A fraction of their rest mass energy, calculated from Einstein's formula $E = mc^2$, appears as energy and the remainder contributes to the inexorable growth of the black hole...

Astronomers have noticed that there are similarities between the phenomena in quasars and radio galaxies. The nuclei of some of the active galaxies strongly resemble mini-quasars. Huge effort is being put into cracking the problem of the mysterious engine generating the energy. The black hole hypothesis is not firmly established, but it does provide a good model for further work.

In the sequence of activity linking quasars and radio galaxies, there are very powerful radio galaxies whose energy, produced in the nucleus, emerges as a pair of oppositely-directed, very narrow, jets extending out to a million light years or so. In some rare cases these jets are dimly visible in the optical region. This proves that the physical processes causing the energy release are located in a central 'factory' that is incredibly tiny by astronomical standards. In a region only a few light weeks across the energy release is the same as a thousand billion suns.

There is a completely different hypothesis which solves the problem of the fantastic energy release in quasars. This alternative view states that quasars are *not* at the distances derived from their redshifts. According to the hypothesis, the quasars are much nearer, and therefore far less powerful. Certain spatial alignments of nearby galaxies and quasars are sufficiently puzzling to give this hypothesis a veneer of credibility. But it probably doesn't apply to all the quasars, the majority of which do seem to be at the distances indicated by the redshifts. There is another problem faced by the hypothesis that the quasars are local: how do they get their observed redshifts? Perhaps they arise by an unknown physical process.

Stars

If we think of galaxies as the dust of the cosmos, the stars for their part are the micro-dust of which galaxies are made. Stars are really very small when we compare them with galaxies: the Sun has a diameter of three light seconds in contrast to the 100 000 light-year span of the Galaxy.

Stars are formed when a part of the interstellar medium condenses. Then, for some reason or other, a region within the gas can attain a much higher density than surrounding regions. The mutual gravitational attraction of the atoms and molecules can become sufficiently strong to trigger collapse about a common centre. Once started, this condensation process gathers momentum quickly, and starts to suck in neighbouring molecules as well. A sphere of gas is formed; energy is released by the infalling material and the shock waves propagating through it. The gas gets very hot, and the sphere starts to emit thermal radiation, at progressively shorter wavelengths as the temperature rises. At the temperature of a laundry iron the radiation is in the infrared. At 1500°C, when iron melts, red light is visible, at 6000°C, the temperature of the Sun, yellow light, and so on, up to millions of degrees, where the radiation is in the X-ray region.

If no new physical process were to come into play, the energy liberated from the proto-star by the gravitational contraction of the gas cloud would be short lived; the star would cool rapidly and become totally insignificant.

But nuclear physics enters the game. And thanks to that, the feeble flicker of gravitational contraction is transformed into enduring stars, with the Sun lasting long enough for life to appear on the Earth. This state of affairs arises because the temperature in the collapsing gas sphere increases enough to break molecules into their constituent atoms, and then to turn the atoms into nuclei and electrons. The result is a very hot gas of nuclei and free electrons. Because of the high temperature, the nuclei acquire very high velocities; when the central temperature exceeds a few million degrees the thermal velocities are big enough to defeat the electrical repulsion between nuclei (which exists because they all have positive electric charge). Then what happens is that when two nuclei collide head on, their nucleons get sufficiently close for the nuclear interactions to come into effect and produce new phenomena, particularly nuclear fusion, which releases energy.

It is quite possible that the first stars were formed a hundred million years after the Big Bang. At that epoch the primordial gas was essentially hydrogen and helium. The first generation of stars was the site of nuclear fusion using protons, which join as pairs in the first step, via the weak interaction, to give deuterium (also called heavy hydrogen, consisting of a neutron and a proton), an electron, and a neutrino.

At the next stage these deuterium nuclei take part in further thermonuclear reactions and produce heavier nuclei. The formation of helium nuclei, or alpha particles, consisting of two neutrons and two protons is a crucial step. These are very stable nuclei, and considerable energy is released when they are made. This is the major source of stellar energy, which has given heat, light, and life to the universe in the last fifteen billion years.

The stars do not just contribute energy. They have also produced the raw material for a variety of objects in the universe: planets, minerals, organic matter,... This is because thermonuclear reactions do not finish with helium. After that, three helium can combine to form carbon, four to make oxygen, and so on, as far as the nuclei of iron, through a complex sequence of strong and weak interactions deep in the superheated cores of stars.

Without carbon, life as we know it cannot exist. The formation of carbon is only possible at all because of an incredible coincidence. For the synthesis to occur the nuclei of carbon need to go through an excited state, that is, be internally agitated, with an energy of 7.224 million electron volts above the ground state. Fred Hoyle predicted on astrophysical grounds that this excited state of the nucleus must exist, and it was subsequently confirmed in the laboratory. If carbon nuclei had had somewhat different properties, the destiny of the cosmos itself would have had another outcome. For physics and metaphysics this coincidence gives us pause for reflection.

How did the carbon and other important atomic nuclei become part of us? This is a consequence of stellar evolution. The life of a star ends when the available nuclear resources are exhausted. There is a relationship between the mass of a star and its rate of energy release which enables us to work out the day of reckoning. The more massive a star, the higher the internal temperature; the higher the temperature, the more intense the nuclear reactions. A star of a few tens of solar masses behaves very extravagantly, burning through its rich energy store very quickly. It is a very hot and luminous star with a very short life. In a million years it is totally exhausted. On the other hand, our Sun, although it has far less matter to start with, is far more economical with the result that it will last for ten or twenty billion years. At the other extreme, a star of one-tenth the mass of the Sun would last for hundreds of billions of years, scarcely hot at all and emitting a feeble glimmer.

In our Galaxy plenty of massive stars have already gone through their ultimate stage. For massive stars this scenario is a huge explosion. As the nuclear combustion flickers out, the star is starved of energy. The interior cools, and then begins to contract. This sets off gravitational collapse whereby the star falls in on itself, or implodes. This sudden collapse releases energy which ignites explosions in a good fraction of the stellar material, and disperses it into interstellar space, along with the precious heavy elements that it contains. This is the phenomenon of a supernova.

Over billions of years the matter in interstellar space gets mixed with the ejecta of exploded stars, and then incorporated into new stars of the second, third,..., and subsequent generations. The Sun was formed 4.5 billion years ago, long after the Big Bang, from material enriched in this way with atoms of carbon and other elements made in stars that have long since disappeared. Those atoms allowed the Earth and all its rich chemistry to feature in the universe.

The successive formation and destruction of generations of stars has intimately controlled the nature and composition of the interstellar gas. Right after the Big Bang it consisted of three parts hydrogen to one part helium. Once the first generation of stars had formed, the galaxies went into a coasting phase; in spiral galaxies around 10% of the matter survives as interstellar gas, less than that in ellipticals, and more in irregulars. A continuous throughput of material was established: stars—gas and gas—stars. This interchange influences the structure of galaxies in a manner that is not properly understood. As a result of this recycling, in hundreds of billions of years the galaxies will contain only cool dwarf stars and gas enriched with nuclei that are so stable that no further energy can be extracted from them by nuclear processes; then even the dwarfs will die away, leaving the cosmos to its distant and uncertain fate.

Clusters of galaxies and filamentary structure

Now that we have examined galaxies in some detail, we are going to stand back and look on them again as the dust of the cosmos, and see if we can determine their distribution in space. This spatial distribution should let us see the structure of the universe itself, just as a map of the distribution of the population in a country would let us see its general shape, the towns, the main highways, and the uninhabited regions.

As we leave our Galaxy on a voyage to explore extragalactic space, we quickly come to two neighbours: the Magellanic Clouds, small satellite galaxies of the Milky Way at 200 000 light years distance. Then we get to the Andromeda galaxy, a wonderful spiral, 2 000 000 light years away. There are others: up to 3 million light years distance some thirty galaxies, mostly pretty small, apart from a magnificent spiral which is visible with binoculars in the constellation Triangulum.

Beyond this there is apparently nothing, not for another 7 million light years. This unusual distribution tells us straightaway that thirty or so of our companions form a group isolated in space, like a village in the countryside. This is the Local Group of galaxies, its two giants, the Andromeda galaxy and the Milky Way, taking under their wing the motley collection of much smaller galaxies, all held together by gravitational forces.

As we push further in our exploration, galaxies start to appear again. Out to 50 million light years they are distributed in about 50 identifiable groups. The nearest of these is in Sculptor and it consists of six fine spirals.

In the constellation Virgo there is a huge assembly of galaxies about 40 million light years away. Instead of the usual few dozen galaxies, this concentration has hundreds in a volume not that much bigger than a normal group. The density of galaxies is therefore remarkably high. In fact, this is the nearest cluster of galaxies, which includes giant ellipticals as well as spirals, and the huge radio galaxy Virgo A with its jet of synchrotron light emerging from the nucleus.

It has been difficult to get this sort of knowledge of the local structure of the universe. We can make a comparison of the relative ease with which stars and galaxies can be studied. With the naked eye you can see 3000 stars and three galaxies. The brightest 1000 galaxies, considered as a set, are around 1000 times fainter than the brightest 1000 stars. From the technological point of view it needs a jump of 1000 times to move from the study of bright stars to bright galaxies. This was achieved in the 1920s with the 2.5-metre telescope (the 100-inch) at Mount Wilson, in southern California. The brilliant discoveries made with that instrument led to the establishment of mountain observatories in the United States. Not until the 1950s did Europeans start to acquire larger telescopes, such as the 1.93-m telescope in Haute Provence (1958), the 2.5-m Isaac Newton telescope (England, 1970), or the 4.2-m William Herschel Telescope (Canary Islands, 1987).

After surveying the thousand or so brightest galaxies, one American astronomer, C. D. Shane, devoted a major part of his life to galaxies a hundred times fainter by counting all detectable galaxies, right to the limit of detectability. This enormous endeavour uncovered a million galaxies. From this number we can immediately reach important conclusions about the composition of the universe. Let me explain why: galaxies a hundred times fainter are, on the average, about ten times further away. This is a consequence of the inverse square law for the fall in observed brightness with distance from the observer. Something twice as far away appears only one-fourth the brightness, and so on. What's interesting about Shane's survey is that galaxies ten times further away fill a volume of space one thousand times larger. And in that much bigger space, the Shane survey found a million galaxies, exactly 1000 times as many as the 1000 in the nearby survey.

The conclusion from this is that out to 500 million light years the population of galaxies is more or less the same as those within 50 million light years. This was a discovery of major importance, but there were others. In his survey, Shane did not find any unusual features about the distribution: no common centre and no preferred direction on the largest scale.

Furthermore, Shane's survey revealed new concentrations of galaxies within the vast uniformity of the cosmos. First, rich clusters containing thousands of galaxies were found, compared with the hundreds in the Virgo cluster. The nearest of these are about 300 million light years away, in the constellations Hercules and Coma Berenices.

The Virgo cluster and all the groups of galaxies in our local neighbourhood, along with the Local Group as well, form a vast system, although it is not quite as big as the Coma cluster. This system has been studied by Gérard de Vaucouleurs, a Frenchman who is at the University of Texas, and he has named it a 'supergalaxy', thus establishing it as a stage in the hierarchical structure of the universe. The centre of the supergalaxy is dominated by the Virgo cluster. The Local Group glides around this, a member of a vast disc system stretching as far as the Sculptor group.

Another American, George Abell, pushed even deeper into space with a special survey designed to find rich clusters and counted 3000 out to 2.5 billion light years. Every one of these 3000 rich clusters contains thousands of galaxies, which in turn are composed of hundreds of billions of stars! Abell also found that the clusters are grouped together into superclusters. The richest of these is an assembly of thirty clusters in a region of space 200 million light years in diameter. This takes the hierarchy up to gigantic systems, the largest entitities found in the universe; the biggest has a mass 10^{16} or even 10^{17} times the mass of the Sun.

Recent research carried out in places as diverse as Hawaii, California, New England, and Estonia has given even more detailed accounts of these superstructures. In addition to the distribution of galaxies and clusters of galaxies across the sky, careful measurements with spectrographs have established the velocities, from which we can deduce the distribution in three dimensions, since the distances are known. Looking at such distributions is a highly subjective process, so the patterns have instead been analysed using computers. They give us a fascinating picture on the largest scale, of some hundreds of millions of light years. Space is like fine lace, a filigree criss-crossed with filaments, chains and strings formed from alignments of galaxies and clusters of galaxies. Between these filaments there are vast empty regions, practically devoid of galaxies.

The filaments are almost one thousand times richer in terms of the density of galaxies within them and yet they occupy only one hundredth of the volume of the universe. In the main the long narrow filaments are made of strings of rich and normal clusters. They apparently correspond to the superclusters discovered by Abell and they are the basic structure governing the large scale distribution of galaxies. Where they intersect they form 'knots', but no one is sure what the significance of them might be. There does not appear to be any thin 'veneer' connecting the filaments.

Generally speaking, it is not easy to give precise meaning to the words group, cluster, and filament, because they merge into each other in a continuous sequence. However, the filamentary structure is a new feature which future theories of galaxy formation must take into account.

Presumably a major event in the early history of the universe determined this outcome.

What happens if we go yet further? On even larger scales, astronomers have made very deep surveys; the number of galaxies becomes so large that they have only been conducted for restricted parts of the sky. All the indications are that the properties found already continue in more or less the same way; no clusters of clusters of clusters have been discovered! We have climbed to the top of the hierarchical structure of the universe. Above scales of about half a billion light years the universe seems to be homogeneous. Very deep surveys of radio galaxies confirm this conclusion. We think, therefore, that as an entity the cosmos is uniformly structured. On the largest scales the distribution of galaxies is essentially the same everywhere, and certainly does not vary from one place to another by more than 50%. These results are endorsed by observations of the cosmological microwave background radiation at 3°K. This fossil relic of the fiery explosion that initialised the universe is essentially isotropic to within 0.1%; it tells us that out to distances well out of reach of the galaxy and radio galaxy surveys, the universe is astonishingly uniform.

This uniformity is the first observational constraint in cosmology. By putting it into the theory of general relativity, it becomes possible to construct model universes giving a good account of the structure and past history of the universe. The second constraint is the recession of the galaxies.

Two great discoveries in cosmology

The recession of the galaxies

The spatial distribution of the most striking cosmic objects, the galaxies, has already shown us something of the grandeur of the universe. If, in our exploration, we include velocities as well as positions, the extra features will add some colour to our story. Consider this analogy: those who have to grapple with traffic problems in a large city aren't simply interested in where the cars are, but in their speeds as well, if only to distinguish moving traffic from stationary vehicles.

However, no galaxies seem to be standing still. All have their own peculiar velocities of a few hundreds of kilometres per second, like bees in a swarm or the molecules of a gas. Furthermore, these speeds are higher if the galaxies are members of a group or cluster. In a rich cluster speeds of a thousand kilometres per second are common. Gravitational forces exerted among and between the galaxies cause these velocities. As they plunge towards the centre of the cluster, the galaxies are accelerated, and then slowed down once they have passed. Rather than the organised orbits traced by planets in the solar system, they follow elongated and inclined orbits through the cluster, just like comets round the Sun. The higher the total mass of a cluster, the higher the average speed of the galaxies within it.

At the present time these motions are hard to disentangle for two reasons. Firstly there is a lack of measurements; secondly, astronomers can only find the radial velocities, those along the line of sight, for which they use the redshifts caused by the Doppler effect. There is no way to find the transverse velocities, across the line of sight, because the corresponding change in position is incredibly small when observed from so far away.

On the other hand, the basic behaviour of velocities on the large scale has been known from the outset of the exploration of the extragalactic universe. There is an overall pattern that swamps the random individual motions. Galaxies are moving away from our vicinity with speeds in direct proportion to their distances from us. This is the Hubble Law. The most distant galaxies that have been photographed with large telescopes are receding at 200 000 kilometres per second, which is 200 times more than the highest internal velocities in clusters, and two-thirds the speed of light.

The recession of galaxies has three characteristics:

1. We see enormous objects, galaxies, each consisting of hundreds of billions of stars, hurtling along at speeds not all

different to the elementary particles, the lightest objects in the universe.

2. This recession is universal, involving all of the visible universe, encompassing billions of light years of distance and billions of years of cosmic history.

3. The law is amazingly simple, just a proportionality. If one galaxy is twice the distance of another, it recedes twice as fast.

Although the relationship itself is very simple, the constant of proportionality, which astronomers refer to as the Hubble Constant, is very difficult to measure. There is nothing quite like a surveyor's chain that we can simply stretch from one galaxy to the next to find the separations. Instead astronomers have to resort to indirect methods, and refining these has kept them busy for nearly half a century. The present consensus is that, to within a 50% accuracy, the value of the Hubble constant is about 20 kilometres per second for every million light years of distance. From this you can see that a galaxy travelling at 2000 kilometres per second would be a hundred million light years away.

The value of the Hubble constant is important for a simple reason. In the past the galaxies were closer together. Backtracking as far as possible, assuming that nothing has changed the velocities in the interim, we conclude that they were touching each other 15 billion years ago. Where did that age come from? It is the time taken to go one million light years travelling at 20 kilometres per second. This discovery is the kernel of the Big Bang theory and the basis for working out when it occurred.

This is a simple argument which we should reinforce with proper theory, and that theory needs to give a correct account of the following deduction: With only a small extrapolation we can imagine a galaxy just one and a half times further away than the furthest that have been photographed to date. Such an object would be receding at 300 000 kilometres per second, the speed of light, and this is physically impossible according to the theory of relativity.

What we need to do is to use that theory fully in order to get a correct treatment of galaxy redshifts.

There is one last point: a galaxy receding at 300 000 kilometres per second would in principle be at a distance 300 000/20 = 15,000 million light years, and we would view it as it was 15 billion years ago, at the moment of the Big Bang. This object would be at the cosmological horizon, which is the distance limit corresponding to recession at the speed of light, and all objects at that distance would be seen in their initial state, right after the Big Bang.

The 3 °K microwave background radiation

If you've ever blown up a bicycle tyre by hand you've probably noticed that the air gets hot as it is compressed through the valve. If we could run time backwards, and reverse the recession of the galaxies into an approach instead, the 'gas of galaxies' filling the universe would warm up. In the limit it would reach such a high temperature that the universe would be filled with a very dense medium consisting of elementary particles with very high energy. These thoughts occurred to George Gamow half a century ago. From them he concluded that the universe had its origin in a Big Bang that was extremely dense and extremely hot; it followed that the present state is the result of 'decompression' from the recession of the galaxies, but that there should be a residue of the initial inferno still weakly warming the universe. A simple calculation in electromagnetic theory enabled him to show that this fossil should be electromagnetic waves of very low temperature.

An analogy perhaps will help us to understand this concept of radiation filling the universe. A furnace used for firing pottery is heated to 1200°C and it is filled with radiation, also having a temperature of 1200°C; this would seriously damage the sight of an operator looking at it without protective eyeglasses. Similarly, when you open the door of a domestic oven heated to 400°C, radiation with a temperature of 400°C flows out, and although you cannot see it you know it is there because of the warm feel of infrared rays. In short, any furnace heated to a given temperature is filled with electromagnetic radiation of the same temperature. The wavelength of these waves is in accordance with Planck's law of 'black' body radiation (and in my opinion this nomenclature is singularly ill-chosen).

The whole universe is a vast furnace. At its creation it had an extraordinarily high temperature which has fallen drastically in the course of its evolution. Gamow made a prediction of -268°C (5°K) for the actual temperature; in that regime the radiation corresponds to millimetre radio waves. However, his prediction was way ahead of the techniques of the time, and was soon forgotten. Around thirty years later Arno Penzias and Robert Wilson, working in New Jersey, accidentally discovered the radiation. The wavelengths of the radiation peaked in intensity in the millimetre part of the spectrum, corresponding to a temperature of -270°C, that is, just three degrees above absolute zero (3°K).

Seen from the Earth, this radiation arrives from all directions in space with amazing isotropy. The radiation has also been studied far out in the Galaxy by using certain types of interstellar molecule as thermometers, and this gives the same result, 3°K. Everything points to the notion that the whole universe is a furnace with a temperature of 3°K, or more properly a freezer given the very low temperature. It is bathed in this cold electromagnetic radiation, left over from a very hot beginning. This fundamental discovery is one of the strongest in favour of the Big Bang theory.

An important consequence is that we have some hard data, namely the value 3°K, which theoretical cosmology can use for modelling the universe. Such theories have progressed so far in the last ten years that we are now able to unfold the past history and thus approach the very first moments of the universe, right up to fractions of a second after the origin. Using general relativity, we can sketch out the big picture of cosmic evolution.

The photons of this fossil radiation constitute the most abundant objects in the universe, and you will recall that they were the first things seen by our imaginary short-sighted parachutist. They are so abundant that on the average there are one billion of them for every nucleon of matter in the universe. They journey through space with remarkable ease, and hardly any of them have ever interacted as they zip along at 300 000 kilometres per second. The last time anything really significant happened to these photons was when they collided with electrons, 300 000 years after the Big Bang. What this means is that for 15 billion minus 300 000 years they have travelled freely, in straight lines, at the speed of light. Those photons received today arrive from close to the cosmological horizon. They are the oldest and most distant messengers we know. For that reason they are very important for deciphering the fate of the universe.

The isotropy is particularly important because it shows that the universe is everywhere the same, even in its most distant reaches. This extrapolates to much greater distances the conclusion about the sameness of the universe already reached from the study of galaxies, radio galaxies, and quasars.

The isotropy is almost perfect. There is only a small deviation of 0.1% which is associated with a particular direction in space, namely the direction in which the solar sytstem in general is moving; this departure from isotropy is due to the Doppler effect. Effectively, displacement in some direction with respect to the general 3°K radiation, which essentially comes from a vast domain of the universe close to the cosmological horizon, decreases the wavelength in that direction and seemingly increases the temperature in the same direction, with a corresponding decrease in the opposite direction. The deviation of 0.1% corresponds to movement at 0.1% the speed of light, that is 300 kilometres per second. Such a velocity is entirely plausible. The 3°K background radiation is thus seen to be a vast reference frame for motion within the universe.

Some radio telescopes, such as the 600-metre Ratan annular telescope in the Soviet Union, have made tremendous efforts to see if there are any tiny irregularities in the nearly uniform distribution of the radiation. If detected, these would correspond to very early stages in the formation of galaxies from the hot gas that filled the primordial universe. According to some scenarios, the collapse of gas should already have been underway, and would have led in succession to condensations on the scale of clusters of galaxies, then galaxies, and then stars. It should be possible to confront these ideas with observations, but so far they have been inconclusive. Such detailed observations would give us insight into the very beginnings of the large scale organisation of matter within our universe.

In conclusion, we see that the universe is immense, but almost empty, sprinkled with a few photons left over from the primordial fireball. Nucleons, the building bricks of matter are a billion times rarer. And yet, against all the odds, stars formed from them, very sparsely distributed but rather numerous, on account of the huge size of the universe: tens of billions of stars in every one of tens of billions of galaxies. And the entire shooting match is whizzing outwards at nearly the speed of light. Can our theories throw some light on this magical country?

4

The relativistic universe

Relativistic space

Four-dimensional spacetime

The twentieth century has given rise to two great theories in physics: relativity and the quantum theory. They gave mankind a radically different view of the nature of the universe. We have to use these ideas in order to understand more clearly the meaning of the quick look at the universe that we ran through in Chapter 3. Relativity particularly has provided a complete and coherent history of the universe from 0.01 seconds after the Big Bang right through to the present age of 15 billion years. When you see the majestic unfolding of this immensely rich tapestry for the first time it takes your breath away.

In seeking perfection we find that Einstein's general theory of relativity replaces Newton's law. In doing so, it replaces the Newtonian gravitational force with a completely different concept: gravitation results from the curvature of space created by masses located in space. This curvature guides the motion of particles, and makes them follow trajectories that correspond to the orbits of Newton's theory. This establishes the general framework for our investigation.

General relativity is a theory of gravitation which followed the results of the special theory of relativity. Some years earlier Einstein had completed the special theory, which is essentially a questioning and redefinition of the nature of space and time. He completely overthrew existing notions and introduced the concept of the four dimensions of spacetime, intimately interwoven, which gave a new description of the real world. We shall use an analogy to get acquainted with spacetime.

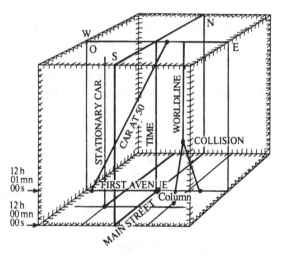

Figure 4. Spacetime of an American town

A map traced on a transparent rectangle represents a small town in the American Midwest with the streets running north–south and the avenues east–west. At the intersection of First Avenue and Main Street is a monument; let's call it Columbus Column. Small filled circles mark the positions of cars today at midday exactly (figure 4).

This transparent plan is a two-dimensional representation of the town at the exact moment of present time. Let us repeat this mapping operation for midday plus 10 seconds, plus 20 seconds, and so on. We make a stack of maps, piled on top of each other, with the successive images of the Columbus Column in the middle and the streets and avenues exactly aligned. The resulting transparent brick is the representation in spacetime of three dimensions of the two dimensional town (figure 4).

Within this transparent set of maps, the positions of the Column define a vertical axis that is a reference for time, irrespective of standpoint, and the intersections of the roads likewise form a spatial grid which is also independent of viewpoint.

What about the traffic? A stationary car traces out a vertical line within the stack of maps, just as the Column does. However, a car clicking along First Avenue at 50 kmph traces out a sloping line (figure 4). The set of lines thus traced out by each car gives its complete history in spacetime within the town. So, if two cars collide at an intersection their two sloping lines in spacetime meet at one unique point; from then on they are a stationary heap of scrap metal represented by a single vertical line.The collision is an 'event'; this terminology has passed into the jargon of relativity theory. An event is a particular point in spacetime. The line traced by a point in spacetime (a car) is a succession of events (the car observed at successive intervals) called the 'worldline' (of the car).

What happens in spacetime diagrams if we set an upper limit to speeds, say 100 kmph for the cars, just as there is a limit in the physical world on the speed of light? The answer is rather simple: the worldline cannot be more steeply inclined than a certain angle. This angle, shown as A in figure 5, is given by a car speeding at 100 kmph. If there is such a speed limit, every worldline is constrained to be less than or equal to the angle A, and they are all contained within a cone with angle A about its vertical axis. In relativity this cone plays a very important role and it is called the 'light cone'. It divides the universe into three different regions: the interior above, the interior below, and the whole exterior (figure 5).

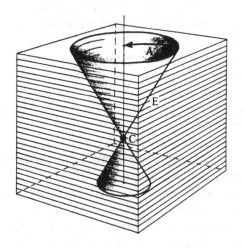

Figure 5. The light cone, C, for the event E

Consider now an imaginary light cone made in a block of transparent plastic, positioned with its apex at some particular point, such as a point E, which we will define as the spacetime location at midday of a car at the Columbus Column. In such a cone, all points inside the upper cone can in principle be reached by the car in the future. It thus defines all locations within the universe accessible to the car from the event E (of course, in this abstraction we are neglecting the fact that houses and fields will be in the way!). In a similar way, all spacetime points in the lower cone were in principle available to the car in the past. Finally, the region outside the cone is the part of the universe that is never accessible to the car. A car at that Column at midday today can never in the future, and could never in the past, reach any place in that domain. This point is absolutely crucial to our understanding the concept of the cosmological horizon.

Actually, there is nothing all that mysterious about this concept. In the analogy, the law sets speed limits. It is clear, for example, that at midday any car travelling along First Avenue and already at Columbus Column cannot come into direct contact with another car that at the same time is somewhere along Main Street. And yet it would be possible to have information, in the future, on what has happened to this particular car we are considering at midday by sending a messenger at (say) 100 kmph to meet it. Therefore the car is not 'contactable' in the sense that it could not also be in Main Street at 12 noon, but it is 'knowable' in the obvious sense that we can get information about it if we so choose. However, we cannot trust this 'thought experiment', because what we have thought through by analogy may be impossible in the curved universe of general relativity. There we encounter events that are always unknowable, a feature that contributes to the frustrating character of the cosmological horizon.

Our imaginary town actually represents the three spatial dimensions of the universe. The Column could be the Earth and the cars the galaxies. The map on a transparent sheet at noon needs to be replaced by a transparent cube showing the distribution of galaxies at noon. We need one such block of plastic for every moment of time. Now how are we going to pile them up? We need another dimension, in addition to the three we have already; but there is no fourth spatial dimension in the real world. In relativity theory there are four dimensions, however: three of space and one of time, but it is impossible for our human senses fully to visualise what this means. Only through mathematics can we get down to a rigorous treatment. If space had only the two dimensions of our imaginary town then everything would be fine. Or would it? In such a flatland we would not be capable of thinking in three dimensions!

The giant leap forward in relativity theory stems from the fact that space and time are treated on an equal basis. This enables the laws of mechanics and electromagnetic theory to be written in a very general form, and gives a simple explanation of otherwise intractable problems. Very prominent among these is the question of the velocity of light, which was addressed in the famous experiments in 1887 of A. A. Michelson and F. W. Morley, exactly 200 years after the publication of Isaac Newton's theory. In an experiment carried out in California, these two physicists showed that the speed of light is always the same no matter what the speed of the observer relative to the light source. If the observer approaches or moves away from the light source, the same velocity is always measured for the light waves, c = 300 000 kilometres per second. The apparent wavelength does change, but not the velocity of light itself.

This revolution in experimental and theoretical physics has extraordinary consequences: lengths contract, mass and energy are one and the same thing ultimately, linked through the famous equation $E = \mathrm{mc}^2$. These consequences transcend purely metaphysical arguments on the nature of space and of time; it is a waste of effort to debate such issues unless you take into account the contributions made by physicists.

Now that spacetime has been introduced, we need to consider the meaning of curvature before we embark on Einstein's general theory of relativity. This is a difficult concept and to make it easier we shall consider only curvature in spaces of the three dimensions, length, width, and height, rather than true four dimensional spacetime.

Curved space

The concept of curved space, or curvature, is very difficult to take on board, although we can give some simple examples. The surface of a billiard ball is, clearly, curved, and we have no problem in accepting that. But to agree that everyday space which we experience all the time is also curved is rather harder to accept. The difference between the two can be perceived thus: the surface of a billiard ball is a two dimensional curved space which we can readily appreciate in three dimensions. To grasp the notion of curvature of three dimensional space we need to think in four dimensions, which is more or less impossible for us. The only sure way to progress is to follow the path carved out by mathematicians such as

Gauss, Lobatchevski and Riemann. In the last century they made curved space an entity to be explored by applied mathematics. The foundation of their work can be perceived with curved space of two dimensions which can be explored from the vantage of the third spatial dimension.

A sheet of paper on a table is a familiar picture of flat two dimensional space. Straight lines, defined as the shortest paths between neighbouring points, enable us to link two points in a way that we would describe as a straight line in everyday language. Such straight lines, known as geodesics, are infinite (casting aside the finite extent of sheets of paper!) in the two directions: one can travel along their own length indefinitely. The space in which they are drawn is said to be open and infinite (figure 6).

A piece of string attached to a drawing pin will trace a circle, with circumference 3.14159...times the diameter. Three points can be joined in a triangle and their three angles will add up to 180°, and so on. This is no surprise. Flat space, drawn on a piece of paper, obeys standard Euclidean geometry, named after Euclid who lived around 300 BC.

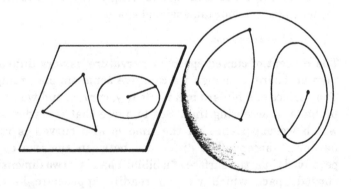

Figure 6. Circles and triangles drawn on Euclidean (left) and spherical (right) spaces

Let us now transfer these exercises from a sheet of paper to the surface of a sphere, and carry them out in exactly the same way. Some surprises result. When a string is stretched between two points on the surface of a sphere the string is constrained to remain in a two-dimensional space defined by the curved surface of the sphere, in other words it has to follow the shape of the globe. Our string thus forms the arc of a circle, in fact a section of a great circle just like the equator or a circle of longitude. This is the 'shortest' path or geodesic, according to the definition of geodesic as the shortest path between two points. However it is not infinite. Any point on the string can be moved along the path and will eventually return to the starting point, having traversed a well-defined distance given by the radius R of the sphere multiplied by 2 and by 3.14159.

We can reach the following conclusion. For two-dimensional inhabitants of this two-dimensional world the space is finite, or closed, but without boundaries. They are living on a spherical two-dimensional space of radius R. For these imaginary beings, who know nothing of the third dimension, a voyage of exploration in the same direction ('straight line') in their space leads to a surprising result, already experienced by those sailors who made the first circumnavigation of the Earth. By continuing along the same track, without deviating to the left, or the right, or above, or below, one eventually gets back to the starting point, which is finally approached from the opposite direction. This outcome, which is in no way unusual from a mathematical point of view, would correspond to our living in a three-dimensional space with a radius of curvature R.

There are other bizarre conclusions. For example, a number of people setting out simultaneously from the same starting point by going in different directions will eventually meet at the 'antipole', or point diametrically opposite their starting point in curved space.

Even if they do not make long voyages, the inhabitants of the surface of a sphere can soon discover that they are not living in a Euclidean space. When they draw a circle on the ground they will find that the circumference is less than 3.14159... times the diameter. Similarly, in drawing triangles they will see that the three angles add up to more than 180°. From such measurements they will conclude that their space is curved and they could measure its radius of curvature. We, for our part, can in principle use very accurate theodolites to make this sort of measurement. In practice the curvature is rather slight, and although the orbit of the planet Mercury displays some features of this curvature, in general we would need triangles with sides several light years in length in order to see the effects of the curvature.

Here is a way of grasping the concepts of the curvature of space. It has the drawback that we shall just consider the two dimensional inhabitants of a two dimensional space, but we shall use mathematical concepts in three dimensions to describe the curvature of two dimensional space. However, our own space is not curved in some supplementary dimension, but rather it is intrinsically curved, which has all sorts of consequences, like having angles larger than we might expect, or smaller circumferences.

But that's not all! Are there spaces for which the opposite holds, where circumferences are larger and angles smaller? Yes there are! You can get an insight into this by considering a saddle surface, rather than a sphere (figure 7). This is an example of hyperbolic space, and it has negative curvature. This is the simplest example: infinite and open.The big question for astronomers is the following: is the space in which the universe exists spherical, or Euclidean, or hyperbolic? The name of the game is to find out if space is finite or infinite. Metaphysically these are fascinating questions. They have stimulated the building of ever larger and more powerful instruments for probing the depths of the universe...

The above is only the simplest aspect of the question. Curved spaces come in an extraordinary range of varieties. Just two examples chosen from two dimensional Euclidean spaces will illustrate this point.

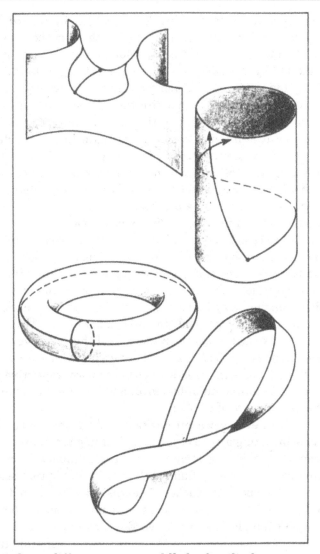

Figure 7. Some different spaces: saddle back, cylinder, torus, and Möbius strip.

Take a sheet of paper on which there are straight lines, circles and triangles, roll it into a cylinder and join together the two edges. From the viewpoint of imaginary beings on this two dimensional space nothing has changed in regard to the properties of triangles and circles, which seem the same as in flat Euclidean space. But consider now a straight line (as perceived by these beings) drawn at right angles to the axis of the cylinder. If we move along its length we eventually return to the starting point from a direction opposite to that in which we set out: in this direction the space is finite. For any other direction it is infinite; seemingly straight-line paths describe spirals (helixes) round the cylinder. Two travellers setting out on straight paths in different directions will endlessly cross each other's tracks, as the two spirals repeatedly intersect. This space is repetitive, like a pattern left by a paint roller; it is Euclidean but infinite in all directions except one.

Here is a case in which the pattern doesn't just repeat along a strip, but goes in two directions, like floor tiles for example. We can start again with a cylinder and make some transformations that are trivial from a mathematical point of view. Imagine a cylindrical tube made of pliable rubber, then join together the two ends to make a car or bicycle tyre inner tube. In two directions on the surface of this torus we can draw trace finite closed paths: at right angles to the tube or parallel to it. Again, this surface is a Euclidean space, repetitive in two directions, finite and unbounded, and much more complicated than the surface of a sphere.

An even more curious case is the Möbius strip. This is made by taking a strip of paper and joining it end to end with a single twist in the strip. As with a cylinder, there is one direction that leads to finite path lengths. But in this case when the starting point is reached, from the opposite side of the strip, left and right are reversed as in a mirror image. This is a simple example of a repetitive Euclidean space which is not

All together, from the viewpoints of general structure and topology, there are five sorts of two dimensional Euclidean space, and 18 varieties of three dimensional Euclidean space. Of these latter, six are finite and orientable, whereas four are finite but not orientable. For three dimensional spaces that have spherical or hyperbolic curvature, the possibilities have not yet been fully explored by mathematicians, except for spaces of constant curvature, that is where the curvature is the same at all points in the space. There is an infinity of varieties; all spherical spaces are finite, but hyperbolic spaces can be finite or infinite.

Thus we can see that the universe may have an extraordinarily rich variety of possible structures. Furthermore, in this book we have thus far only considered three dimensional spaces. When spacetime in four dimensions is tackled using general relativity the variety becomes frightening. We are at once in a jungle that goes beyond the inventions even of science fiction! The wonderful flexibility of relativity theory also enables us to understand things as diverse as the expansion of the universe and the astonishing geometry of black holes.

General relativity

Einstein's equations

Newton's theory tells us that a planet is held in its orbit round the Sun by a force of attraction that extends across the distance between the two bodies, and thus makes the planet move as if a piece of elastic were stretched between the two objects. This wonderful theory gives an account of planetary motion that is remarkably precise.

However, Newtonian theory fails in one tiny detail associated with Mercury's orbit. This planet, the nearest to the Sun, has a rather elliptical orbit, with the major axis of the ellipse pointing in a certain direction.. The combined gravitational pulls of the other planets cause this major axis slowly to alter its direction in space. When this effect is analysed using Newton's theory there is a small discrepancy between the theory and the observations, equivalent to the major axis making one extra turn over a period of three million years.

The general theory of relativity, however, gives a perfect account of the anomaly. This was a triumph for the theory; although the point at issue seems to be very small, science has to be ceaselessly alert to small effects. In the last 70 years since this success, more and more refined tests have confirmed relativity theory. Moreover, remarkable, often unexpected and revolutionary, predictions have emerged from the theory, and yet they have been verified in nature. It is for these reasons that relativity theory has a central place in modern science, not just in terms of applications, but also for what its fundamental conceptions imply about the universe.

General relativity gives a different meaning to gravitation than does Newton's theory. A simple model will give us a foretaste of what is to come. Imagine, once more, that we are in a two dimensional space represented by a sheet of rubber stretched over a horizontal frame. This is a non-curved Euclidean space. In the middle, place a heavy ball to represent the Sun; the rubber stretches, of course, and the ball then sits in a wide hollow. This hollow represents the curvature of space which the Sun's mass causes (figure 8).

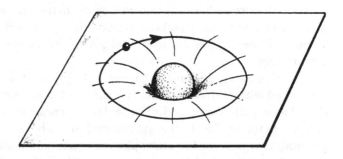

Figure 8. A model of the gravitational interaction between the Sun and a planet, according to the general theory of relativity

Now let a small ball be a planet going round the Sun. It can be projected tangentially around the hole, with an adequate velocity, and it will then travel in a circle round the side of the hollow. This model shows us how gravitation works according to general relativity: the Sun curves the space in its vicinity, and a planet has to follow the contours of that curvature.

This simplified picture enables us to formulate the essentials of Einstein's theory:

1. Four dimensional spacetime is a curved space in which the amount of curvature is determined by the masses in the space.

2. The world lines of masses in this curved spacetime are geodesics (We ignore non-gravitational interactions at this point).

These statements use the jargon of the theory, but we can express them more easily although less precisely:

1. Both space and time, which are intimately linked, are affected by the presence of massive bodies.

2. A body moves in space, as a function of time, in such a manner that in spacetime its track is the shortest path (this is the definition of the geodesic), and this exactly follows the curvature.

Since the advent of Einstein's theory, these statements are not formulated in words, because prose is neither precise enough nor complex enough to take us very far. The complexity of spacetime is handled using geometric theories established by Riemann. He used tensors, a sort of super-vector in spacetime with 16 components (although vectors in three dimensional space have only three components), to represent the geometry of spacetime. Massive objects are specified not just in terms of position, but also in terms of their masses and velocities by another tensor, the energy-momentum tensor. Finally, the relationship between the curvature of spacetime and the distribution of massive objects is expressed by means of equations linking these two tensors: the famous Einstein equations of general relativity.

This system of 16 equations is interrelated in such a way that nobody has ever managed to find the general solution to them; the problem is so intractable that it is perhaps beyond the capacity of the brain of *Homo sapiens*, although that hasn't stopped people from trying! Fortunately for us, on the grand scale the universe is homogeneous and isotropic, to within a

50% certainty. This leads to considerable simplification, and the Einstein equations can then be used rigorously to represent the behaviour and evolution of the universe at large. Such solutions are the foundation of our present relativistic models of the universe.

Relativistic model universes

When gravitational theory, according to general relativity, is applied to the universe at large, there are simplifications arising directly from the isotropy and homogeneity on a length scale larger than half a billion light years. This is the typical scale size associated with the filamentary structure seen in the distribution of galaxies, clusters of galaxies, and superclusters.

Under these conditions spacetime, as described by Einstein's equations, is split up into two parts which can be treated separately:

1. A three dimensional curved space with the same radius of curvature at every point.

2. A simple equation describing the variation of this radius of curvature as a function of time.

The nub of the simplification is this splitting into separate components. The mathematical simplicity arises from the homogeneity and isotropy of the contents of the universe. Three dimensional space corresponds to the space that the universe lives in. According to the sign of the curvature this space is spherical for positive curvature, hyperbolic for negative curvature, or Euclidean for the half-way house of zero curvature.

The time component reduces to a universal time or 'cosmic time'. Its universal properties are a result of the homogeneity of the universe because general relativity as such does not recognise any concept of absolute time. Indeed, special relativity introduces the concepts of time dilation and the non-conservation of simultaneity. The privileged position of cosmic time is similar to the 3°K universal microwave background radiation which provides a quasi-absolute reference frame, in spite of length contraction, embracing a vast homogeneous zone of the cosmos. What exactly is cosmic time? Suppose there were a vast transgalactic airline system: cosmic time is the time that would be registered by clocks in every airport, having been synchronised at some point by means of radio signals.

Once the spacetime equations have been separated into equations for space on the one hand and time on the other, the simple equation describing the behaviour of the radius of curvature as a function of time leads to two possibilities:

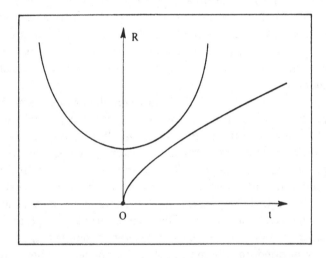

Figure 9. Two cases of the expanding universe, with and without a Big Bang.

1. In the infinitely remote past the universe had an infinite radius of curvature but this decreased progressively in the course of time to reach a certain minimum (and therefore a maximum amount of curvature). Thence it rebounded, and will expand indefinitely into the far future (figure 9). Observationally this model is rejected because it is inconsistent with the detection of the 3°K microwave background, which teaches us that the universe was once fantastically compact.

2. At some point in the past the radius was zero (infinite curvature) and from this state it expanded extremely rapidly at first, followed by a more gradual expansion. Observationally this case is rather attractive for two reasons: firstly, it is consistent with the 3°K microwave background; and, secondly, the recession of the galaxies can be explained via the expansion of space.

In fact, this second case is the basis of the Big Bang theory. Remarkably, Friedmann had already found these solutions to Einstein's equations in advance of Hubble's discovery of the recession of galaxies. At that time the very idea of the expansion of space seemed so alien that Einstein himself modified his equations to avoid such an eventuality by introducing a then artificial term known as the 'cosmological constant'.

At this juncture it is worth emphasising a point that the public at large finds rather confusing: fitting the observed recession of galaxies into an expanding universe. Generally speaking we imagine galaxies scattering themselves into some pre-existing space, like the shrapnel of a bomb. In fact, the relativistic interpretation shows this to be incorrect. Instead we must think of the galaxies like currants in a cake that expands progressively as it bakes. That is a different concept, and seems contrary to common sense. However, much garbage has been propounded in the name of common sense about general relativity in the first half of the twentieth century!

As space grows, it carries the galaxies and their stars with it, locked into the general expansion. This begs the question: if space is expanding, into what is it expanding? Isn't there somewhere that it spills over? The answer is no. To grasp this we must once again return to two dimensional space. The easiest sort to understand is the spherical space of the surface of a sphere. Now, if the radius of curvature of this space increases along the third dimension, then from the point of view of imaginary inhabitants of the surface the space in which they live increases in extent. For them this is an important geometrical consequence, although the properties of three dimensional spheres might be quite insignificant to them, whereas for us it is a feature of the model. We ourselves live in three dimensional spherical space, and for us the equivalent is the increase in the curvature radius with cosmic time. The volume of the universe increases without it cascading into an inaccessible fourth dimension (which would only be accessible to four-dimensional beings).

All of this is hard to grasp but mathematically it is completely rigorous. Indeed, if space had been curved to such an extent that the ancient Egyptians could have detected it from their surveys of fields by the Nile (by seeing that the sum of the angles in a triangle was more than 180°), then Euclid would probably have invented Riemannian geometry straight off! And then common sense would have said that Euclidean geometry is a load of nonsense.

In the second case presented above, the expanding Big Bang model, Einstein's equations have two possible outcomes:

2a. Either, progressively as cosmic time passes the expansion continues indefinitely (figure 10);

2b. Or, it steadily increases, reaches a maximum, then starts to contract in a manner that mirrors the expansion, until it eventually collapses into the nothingness from whence it came.

Both possible outcomes for the future of the universe have fundamental consquences for its ultimate fate, and they are of profound metaphysical importance. The choice between these two is of immense significance and is a major objective of the Hubble Space Telescope and other large aperture telescopes that will operate from the ground.

For the moment our historical sketch of the universe, underpinned by general relativity and observations of the distribution of matter in the universe, of redshifts, and of the microwave background, has two scenarios: the Big Bang leads to an expansion which continues forever, or it turns into a contraction and eventual extinction.

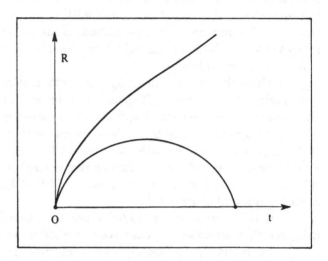

Figure 10. Expansion in the Big Bang, showing the effects of indefinite expansion, and of eventual contraction.

So far as the curvature is concerned, each of the above two cases has two further possibilities: spherical space (which would have a finite volume) and hyperbolic space (which can be finite or infinite in volume). Theoretical possibilities have been greatly reduced by the observational data in order to end up with this small number of cases. Now the observational effort is directed at getting even more precise data which will discriminate between these possibilities and find out what kind of universe we live in. Let's now look at the panorama of the universe as we understand it.

The panorama of the universe: a 'Grandiose Frescoe'

Theory gives us equations and observation yields numerical data. Let us now go back in time to within one second of the primordial fireball. Why the first second in particular? Compared to the age of the cosmos one second is ridiculously small. However, in the initial holocaust which determined from the outset all subsequent events in the universe, the first swing of the cosmic pendulum was of major importance. The situation is analogous to a thermonuclear explosion; physical processes in the first few instants of time decisively determine all the effects of the explosion; in the case of the universe itself the initial violent explosion is unimaginably larger.

One second after the Big Bang the universe had a meaningful temperature: 10 billion degrees, everywhere. At such a temperature no molecule, atom, or even atomic nucleus can exist. The matter exists only as elementary particles: protons, electrons, and neutrons. Because space is fantastically compact at this stage the material density is 10^{-4} gram/cm^3, which is enormous compared to the present value of 10^{-31} gram/cm^3, but not at all unusual as far as physics is concerned. In a neutron star, for example, the density approaches 10^{15} gram/cm^3.

The extremely high temperature, on the other hand, leads to a very high equivalent density for radiation, which consists mainly of gamma rays. They contribute, through the energy-matter equation $E = mc^2$, a density of 1000 gram/cm^3, which exceeds the matter contribution by ten million times. It is hard for us to imagine this situation in which the density of the universe derives to such a huge extent from radiation. That is one of the surprises of the early universe, and the situation prevailed thus for a long time.

When the universe was ten times older (or should one say ten times less young?), and the cosmic time registered 10 seconds, the headlong expansion of space had already reduced the temperature to a few billion degrees only. Among the effects this had was this: the radiation became less energetic and, at the same time whenever a neutron and proton collided the resulting deuterium nucleus had a smaller chance of being smashed apart by the weaker radiation. So the cosmos became a deuterium factory, in which production peaked after three minutes. Subsequently the production switched to helium 3 (made from one deuterium nucleus and one proton) and helium 4 (two deuterium nuclei), with a trace of the next element lithium (formed from three protons and four neutrons).

After about a quarter of an hour nuclear reactions ceased for two reasons: by then the temperature had fallen to a few hundred million degrees, and that is not enough for individual nuclei to beat electrical repulsion and make contact; also, the available space had become much larger — between one second and a quarter of an hour it increased by one million times — so that the nuclei had more or less no opportunity to meet in those more dilute circumstances. So, after 15 minutes the net result of the nuclear reactions had been to make a cosmic soup consisting of three-quarters protons, one-quarter helium, together with a little deuterium and a trace of lithium.

Together with the electrons, these various nuclei made a very high temperature gas, bathed in powerful electromagnetic radiation, now comprising principally of X-rays, which are less energetic than the gamma rays that dominated in the first second. The particles and radiation together made an intensely bright plasma. At the prevailing densities, the electrons prevented the radiation travelling any significant distance because strong interactions between electrons and photons made the plasma opaque, despite its being filled with radiation. Such interactions ensured that the temperatures of the matter and the radiation were equal, a state that astrophysicists term local thermal equilibrium.

The expansion continued unabated, and the temperature of the particle-radiation mix continued to decline. For many centuries nothing much happened, in marked contrast to the fast moving action of the first fifteen minutes. Progressively the temperature fell, and in parallel the plasma became less dense and less brilliant, more like thick fog; the radiation passed from the X-ray region to the ultraviolet.

At an age of 300,000 years the temperature reached 5000°K, and at that stage the radiation could no longer break up associations between electrons and nuclei formed in particle collisions. Thus the first atoms came into being, atoms of hydrogen and helium. Matter changed suddenly from being ionised (electrons not attached to any nuclei) to being neutral (electrons being members of neutral atoms). Once tied down to atomic nuclei, the electrons could no longer impede the radiation. This is called the recombination era, at which matter and radiation became decoupled.

This is a crucially important moment in the evolution of the cosmos because from this time radiation travels unhindered. The consequences of this are important for the 3°K background radiation: what we detect right now has propagated freely for 15 billion minus 300,000 years, which is to say that it reaches us from regions located 15 billion minus 300,000 light years away. Further than that it is lost in the opaque plasma filling the universe up to 300,000 years after its birth. Subsequent to that time it has made its long journey to our radio telescopes, continually falling in temperature because of the expansion, until it has reached the 3°K of the present epoch.

Around the same time, 300,000 years ago, radiation finally lost its supremacy over matter as measured by its contribution to the density of the universe. Matter, seemingly so insignificant to start with, then asserted itself through its gravitational force, and in fact allowed expansion to proceed a little faster.

After recombination, nothing of much interest happened for tens of millions of years; profound boredom gripped the whole cosmos. The decline of temperature and matter density continued of course, and the gas became less and less dense, as well as progressively colder. The colour of the radiation changed from white hot, to yellow, orange, red-hot, and then a feeble reddish glow... After about 10 million years it reached 20°C, which we would have found rather comfortable had we been there! But that didn't last long, and after 100 million years it had plunged to 200°C below freezing, with a density of 10^{-27} gram/cm^3, which is around one hydrogen atom per litre, just like interstellar space within galaxies today.

At this point one might think the universe should have reached the end of its career, which would have been a disappointing outcome of such a dynamic beginning, and no people would have been around to wonder about its purpose. In fact, after 100 million years of this characterless existence galaxies started to form by sheer chance. This is only a hypothesis, although it is a reasonable one: in the ubiquitous medium filling space there were, here and there, parts more dense than the average. By means of a process similar to star formation in galaxies, these local condensations gradually grew in influence, and they ultimately turned into galaxies. Nobody knows whether stars formed first, then gathered into galaxies and clusters, or whether superclusters came first, and then fragmented into galaxies, which in turn permitted star formation. But this is the stuff of professional astronomy: theorists have numerous models but the observers have almost no data.

There is one positive bit of information: at an age of 1 or 2 billion years the quasars already existed, and therefore we can guess that there were probably galaxies and stars too. Therefore, by 1 or 2 billion years the destiny of the universe had reached a decisive turning point. The slow decline of its fantastic primordial energy was substituted by nuclear energy liberated in stellar interiors, energy which, fifteen billion years later still brings warmth and life to a universe so majestically dispersed through the vastness of space.

Once galaxies had formed, the rest of the time was used up with successive cycles of star birth and star death within galaxies. Within our Galaxy the Sun and Earth formed 10 billion years after the origin, 4.5 billion years ago. After that there was a further blank period lasting a billion years before the appearance of primitive lifeforms. Then *Homo* appeared some millions of years ago, after which events of increasing cultural importance took place in an accelerating succession.

After this there are two possibilities: either the expansion continues indefinitely or it is reversed in a final implosion. Neither outcome is particularly attractive from our point of view! In the first case, the temperature will gradually approach absolute zero, -273°C. As far as radiation is concerned the consequences are not immediately serious. It reaches an endpoint only when there are no more stars left with nuclei that can go through nuclear fusion, and that will be

27 mai 1985, 21:20 T.U.   153 mm   280X   J. H.

The lunar crater Manilius

The Andromeda Galaxy

The Whirlpool Galaxy in Canes Venatici

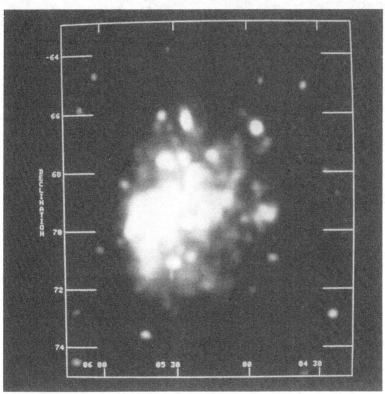

An image of the Large Magellanic Cloud in radio waves

The telescopes in Hawaii

The Very Large Array radio telescope

The Horsehead Nebula

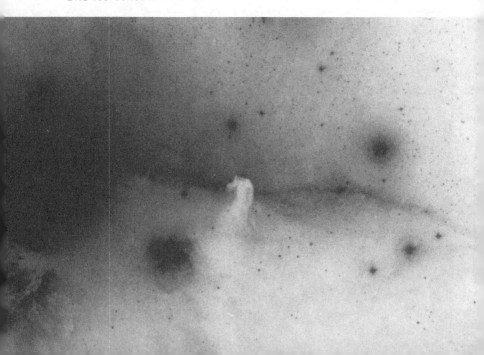

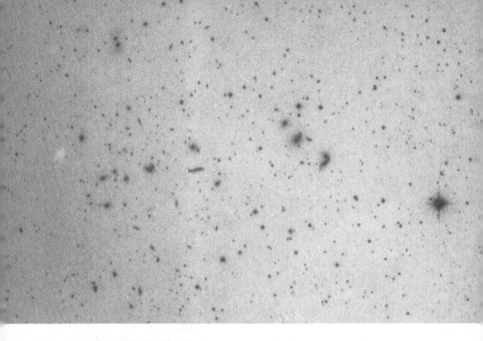

Galaxies of the Hercules Cluster

Survey of one million galaxies shows the filamentary structure of the universe

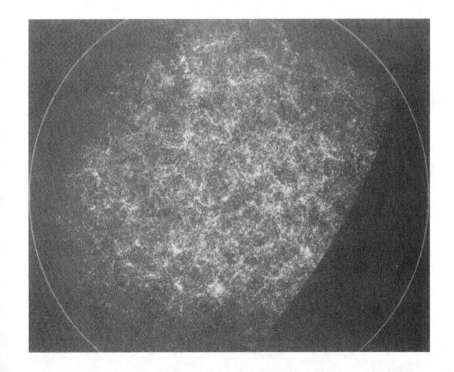

The Hubble Space Telescope

Dried riverbed on the planet Mars

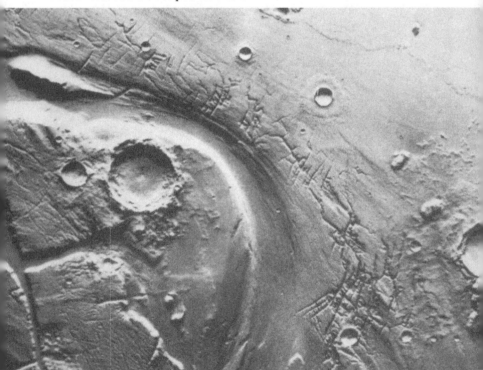

Saturn and its satellites

Footprints on the Moon and from Australopithecus in Tanzania are separated by three million years.

hundreds of billions of years in the future for the most economical stars, although an ultimate day of reckoning will come even to these. While there are stars there will be energy, and heat, and life. Galaxies will get spread out more and more, our own will be all alone, isolated in space that becomes emptier and emptier. This is death from absolute cold...

It is possible that a civilisation could be astute enough to find some means of producing energy. But even if new technology is invented, perhaps allowing energy to be extracted through black holes, can there be indefinite reprieve?

Let's now look at what fate awaits the universe if it takes the other track. The universe expands more and more, pauses, and then shrinks, to fall back in on itself in a symmetrical reversal of the expansion. Since the universe is already 15 billion years old, the slowdown and then reversal must both take at least as long as that. Altogether the universe would probably exist in a state we would find acceptable for more than 50 billion years. Throughout it would have stars and 'reasonable' temperatures. Even so, there is an endpoint, in the final Big Crunch. Just as there were 10 million years from the Big Bang to a temperature of 20°C, so in the collapse the same time will pass as the temperature goes from 20°C to the inferno of 10 billion degrees, just one second before the total collapse of space. Is there nothing that can be done?

As a matter of fact, cosmological solutions to Einstein's equations allow the contraction phase to rebound in a new expansion. The universe would thus experience a succession of cycles of expansion followed by contraction followed by expansion... For such a situation, a detailed examination of the physics shows that the critical, extremely hot, phase lasts only a short length of time: about one hour above a temperature of 100 million degrees. Technologically the problem of saving civilisation is not insurmountable; plasma physicists working on controlled nuclear fusion can already manipulate magnetic fields that are capable of holding plasma at such a temperature. If a super-civilisation, fifty billion years in the future, manages to construct an anti-collapse machine which will last for an hour, would they emerge in the following cycle of the universe?

Such a civilisation would have the marked advantage of beginning very soon in the cycle, instead of having to wait ten billion years for evolution to commence. But, of course, these are mere speculations, given our present state of knowledge.

Presently, the fate of the universe, hundreds of billions of years into the future, has two options (excluding for now the instantaneous and unpredictable catastrophe suggested by the theory of the inflationary universe). Which of these two destinies has been chosen?

Unfortunately observations are not precise enough to help us decide the issue right now, although they have some slight tendency to favour indefinite expansion. In order to make further progress it is essential to measure the rate at which the expansion is decelerating. To measure the deceleration parameter, one of the most direct methods involves seeing what the speed of recession was in the past, by measurements in deep space, because then the light travel time lets us see into the past. This is a major initiative behind the Hubble Space Telescope which NASA will launch with the Shuttle in late-1989. Orbiting above the atmosphere, this telescope will have a clear view of the far universe.

The cosmological horizon

With this general picture of the history of the universe, we can now take another look at the cosmological horizon, which has been mentioned several times. Then, having journeyed so far in space and time, we shall close our relativistic window on the universe; instead we shall turn inwards to the microstructure of the cosmos and, in doing so touch on quantum physics.

Once again we need to use the two-dimensional model of space that we made by piling transparent maps on on top of the other. Columbus Column will represent our Galaxy. Maps going back almost to the Big Bang, 15 billion years ago, record the passage of all the photons we receive today. Their worldlines are inclined at an angle which corresponds to their velocity of 300,000 km/sec, and they all intersect at the point E in spacetime (the event), representing our Galaxy right now.

The world lines follow the surface of an inverted cone with its apex at point E, and successive slices through this cone take us right to its base, which corresponds to the Big Bang. The side of the cone defines a circle on the base; this is the cosmological horizon (figure 11).

Effectively, all points in three dimensional spacetime that are located inside the light cone represent events that we could contact. In particular, all points of two dimensional space enclosed by the circumference are contactable; to express this another way, all these points could send us information by messenger. A factor of special importance concerns points actually on the circumference: they are the most distant locations that can in principle be contacted and information from them has to travel by the fastest possible couriers, which are photons. We can summarise thus: all points in space within the circumference constitute the observable universe today, those located beyond it are unobservable, and those on the circumference form the cosmological horizon.

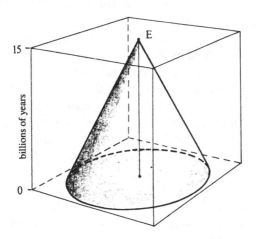

Figure 11. The cosmological horizon for the event E

The cosmological horizon effectively draws its existence from the fact that our universe had a beginning and information cannot circulate inside the universe any faster than the speed of light. The most remote objects, furthermore, are on the cosmological horizon and we see them in the state they had at the moment of the Big Bang.

Are we saying here that there is nothing beyond the horizon? Surely not. As time flows on, what will happen? Consider what happens as the point of spacetime representing our Galaxy (the event E) moves up in time through the layers in our transparent model of spacetime. The light cone will expand as it travels up to higher layers, and the circumference on the bottom layer will grow in radius, thus encompassing more and more space. The cosmological horizon expands as time passes. Now, we must assume that there is nothing particularly special about the present moment of time relative to the vast history of the universe; so it is almost certain that parts of the universe which will be observable next week (for example), already exist but they are unobservable right now.

However, we must not think that these thoughts, applied just to a small time interval like a week, can be extrapolated to much longer time intervals. Curvature, which general relativity tells us is created in spacetime by mass, leads to bizarre situations. Light cones, for example, keep their basic properties, but the curvature causes them to have curved sides (figure 12). There are examples of relativistic model universes in which the apex of a light cone can extend indefinitely into the future while the circumference of the base gets fixed in radius, just as if it is glued down to space. In such a model universe no matter how far into the future one cares to go, it will always only be possible to see a limited region of space; certainly that would be a frustrating kind of universe to be in!

I cannot resist describing one extraordinary phenomenon in these frustrating universes. We shall imagine that a star is observable just at the edge of the horizon, and it is therefore at a point on the circumference defined by the light cone today (figure 12). In a week's time (say), the light cone will, as it moves higher, cut the world line of the star at some

layer a little higher up; the early history of that star will be
revealed at that point. However, if the apex of the cone now
rises indefinitely but the base is fixed at some limiting
circumference then the sides of the cone will never cut the
world line of the star at any higher layers (that is, at any later
times). From this it is clear that even in the infinite future there
is no way of knowing more than a short portion of the early
history of that star; nobody will be able to see its entire history.
Furthermore, as time passes its observed behaviour will
proceed more and more slowly as it gradually approaches the
end of the section we can follow, rather like a film or video
running slower and slower. This is a rather spectacular
example of the effects of time dilation, in the case due to a
complex cosmological feature of spacetime curvature.

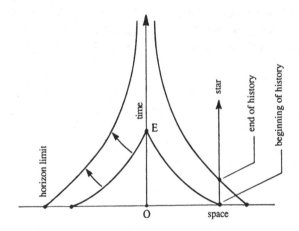

Figure 12. This model universe has a doubly frustrating
cosmological horizon: only the early history of a limited part of
space can be known

From this illustration we see that universes can exist in which it is not only impossible to see all of space, but it is also impossible to find out about all of the history of the accessible volume. This is a doubly frustrating case! The above raises a question that is interesting, at least from the metaphysical viewpoint: are we prevented, in such cases, from including regions beyond the observable in our reckoning? That will all depend on the precision with which we can make measurements in the region that is accessible and on the effectiveness of the theories used in interpretation.

Such a possibility is rather encouraging and gives an incentive to study the universe. But precision is always circumscribed by technical or quantum limitations, so we must be cautious and keep our extrapolations within reasonable bounds.

An example of that is given by the biggest question of all: 'What was there before the Big Bang?' By adding quantum aspects to the relativistic picture of the Big Bang we can attack more fully problems connected with the supercondensed phase at time zero, as well as start to formulate some preliminary conclusions about its random or 'probablistic' character.

5
The quantum universe

Quantum uncertainty

Spin

I want to start this section with a description of particle spin. In
addition to the usual properties of mass, charge, position and
velocity, particles have a further parameter which describes
their state of rotation. Physicists call this 'spin'. We have
already seen that in quantum physics it is impossible to
measure simultaneously, in a precise way, both speed and
position (recall the Heisenberg uncertainty relations). The
same considerations apply to spin, which cannot be absolutely
determined with arbitrary precision. The things that we can in
principle find out are the rate of rotation (that is, the number of
revolutions per second) and the component of this rotation in a
given direction in space (for example, the angle of the axis of
rotation relative to the direction of motion).

It is important to understand that these two knowable
quantities can take only certain quantised values. This is an
aspect of quantum physics that we have already encountered
when describing wave motion. We can use as an analogy the
stable vibrations that can be sustained on a piano string: the

basic note is generated when the centre of the string vibrates to its maximum extent while the two ends are (necessarily) fixed. This is the fundamental mode of vibration and the wavelength of the wave is twice the length of the string. We call this the fundamental frequency of vibration. There are other possible frequencies of vibration which are exact multiples of the fundamental; these are called the harmonics and they are generated when the string vibrates in an exact number (2, 3, 4...) of sections. The harmonics are an exact (quantised) multiple of the fundamental.

The same considerations apply to spin. There is a fundamental base which is directly related to Planck's constant. In this picture the electron has spin 1/2; relative to the direction of motion, this spin is either -1/2 or +1/2. The first case corresponds to left-handed or negative spin and the second to right-handed or positive spin. The photon spin has the value 1, and it is expressed as -1 or +1 according to direction.

Particles with half spin are fermions and those with integer values of spin (including zero spin) are bosons. These two categories emerge also because, collectively, their statistical behaviour is different. Fermions *en masse* obey statistical patterns first set out by Fermi (hence fermion), whereas the second category statistically follows a formalism developed by Bose (hence the name boson).

Quantum fuzziness

The universe is one immense entity composed of miniscule elements; its destiny depends on their intimate properties as much as on its gross structure. The same is true in the everyday world. Two knives, one of mild steel and one of stainless steel, might look the same and perform equally well on the large scale; on the molecular level they are not the same, however, and the stainless steel one will last much longer, because it will not corrode. We now want to turn our attention to the universe on the microscopic level, and in this world quantum physics reigns supreme.

The feel of this type of physics can be conveyed by the expression quantum fuzziness. Here once again, as with relativity, common sense is deceiving. But quantum theory works very well, extremely well in fact, so we have to use it. It will take us into situations that surpass even the craziest science fiction.

Heisenberg's uncertainty principle expresses quantum fuzziness in terms of mathematical physics. This is somewhat difficult to grasp, unfortunately, but nevertheless we shall have to refer to it several times.

Let's consider an elementary particle, such as an electron. It is impossible to determine precisely, at one and the same time, both its position and its velocity. If the position is well defined then the velocity will have a large uncertainty. Conversely, if we know the velocity very accurately, it is impossible to say very precisely where the particle is. The more we know about one, the less we can know (in principle and in practice) about the other. Another way of expressing this is to say that uncertainty in position varies inversely with uncertainty in velocity, and that uncertainty in velocity varies in inverse proportion to uncertainty in position. Most importantly, the uncertainty in position multiplied by the uncertainty in velocity is a universal constant. This quantity is the Planck constant, h, which is of as great an importance in physics as are the velocity of light, the charge on the electron, and the gravitational constant.

This constant and the uncertainty relation apply to all particles in a way that depends on their masses. In the correct expression of the uncertainty principle it is not just the velocity as such that counts, but rather the velocity multiplied by the mass, that is to say the momentum of the particle. For example, a marble with a speed of 2 centimetres per second and a mass of 1 gram has exactly the same momentum as a 2 gram marble travelling at 1 centimetre per second. Momentum measures the effort needed to resist the motion of the marble.

Heisenberg's uncertainty relation can be written like this:

uncertainty of position times *uncertainty in momentum* = *h*.

Two important statements can now be made: *h* is a very small number because the uncertainties are actually very tiny. Furthermore, for objects with significant mass, like marbles, the uncertainty in their momentum is very small indeed, and therefore the corresponding uncertainty in velocity will be just as small. Clearly, in the everyday domain of the real world (marbles, people, planets,...) the intrinsic imprecision that flows from Heisenberg's relation is so small that for all practical purposes we can measure position and velocity with enormous precision. In other words, happily, Heisenberg's uncertainty principle does not interfere with everyday life, and therefore with the real world as we per-ceive it.

The smaller the mass of a particle (think of an electron now) the more the uncertainties come into play. They are not negligible at this level: suppose you know an electron's velocity to within 1 centimetre per second, then it is impossible to know its position to within better than 1 centimetre. For a proton, which is 2000 times more massive, the same calculation gives an uncertainty in position of 5 microns.

It is not fundamentally difficult to use Heisenberg's uncertainty relations, but it is hard to conceive of their reality. Perhaps an analogy can help with the visualisation. Imagine an electron trapped in a tiny box. Now make the box smaller; as we do so the electron's speed will have some value between zero and that particular value indicated by the uncertainty principle, which will increase as the box shrinks. An electron constrained within a 1 centimetre box will necessarily have a velocity somewhere between 0 and 1 centimetre per second. A rather graphic picture, but a misleading one in some ways, is to think of a fly trapped in your two hands. The tighter you squeeze them together the faster it will buzz around in desperation!

This Heisenberg relation is concerned with spatial data: positions and speeds. There is a further relation which with temporal information, and the measurement of energy. It arises from quantum physics and was immediately assimilated into special relativity and its underlying spacetime. According to the time relation, any uncertainty in the energy of a particle is inversely proportional to the uncertainty of the actual time during which the measurement was made, and both are linked through Planck's constant.

A really important consequence of this second relation touches the sacrosanct principle of conservation of energy. It is possible for it to be violated and the degree to which that occurs will depend on the time for which it lasts. In the limiting case of the real world, measurements take place over sufficiently long time intervals for energy to be more or less perfectly conserved. But for very short time intervals significant fluctuations in energy can occur. So, the energy fluctuations of an electron are comparable to the rest mass energy ($E = mc^2$) for times of 10^{-20} seconds.

There are major consequence flowing from this. For example, with a probability of once every 10^{-20} second an electron in principle has at its disposal enough energy to create a further electron. On the average, however, order is imposed on this chance event, though later on it could still happen according to purely random fluctuations.

Heisenberg's uncertainty relations have a major influence on physics and therefore on the universe. They are essentially tied into the wave-particle behaviour of both elementary particles and radiation, and through them receive experimental confirmation, which shows that despite their bizarre aspects we have to believe them.

Waves and particles

Until the beginning of the twentieth century, light had all the appearances of a wave motion propagating through space. Just like waves on the surface of a liquid, they have wavelength (the distance from one wave crest to the next), velocity, and frequency (the number of waves passing per second). They can reinforce each other and interfere. Two waves with the same length excited in unison from two different points create standing waves. By combining one wave with several of its harmonics (waves with twice, three times, and so on, the basic frequency) it is possible to produce waves with a step profile, such as tidal bores.

At the beginning of the twentieth century these views were challenged: Planck and Einstein discovered that light waves can behave like particles. In interactions with matter, light waves can only transfer energy in packages of a certain amount. The energy of these packets, or quanta, is governed by the following relation:

energy = Planck's constant times *frequency*

The energy of a wave which seems to be distributed throughout the vibration is actually only manifest through the quantum aspect, that is to say as photons or individual packets of light.

In addition to this, electrons, which have all the appearances of elementary particles, do also have a wave aspect. For example, if a beam of electrons is directed at a target with two small holes in it, behind which there is a screen, the electrons passing through the pair of holes produce interference fringes just as if they were waves. This wave aspect of electrons is exploited in the electron microscope, in which the imaging properties of waves are achieved by particles instead.

In the 1920s, Louis de Broglie established a theoretical framework for explaining this puzzling behaviour: the theory of wave mechanics. In this scenario one associates a wave with every particle, the wavelength of which is given by Planck's constant divided by the particle's momentum. Physical phenomena are then interpreted by saying that the wave gives the particle position in a probablistic fashion: where the wave is strongest you have the highest chance of finding the actual particle. The wave is not a material wave, but a probability wave. However, this structure, whose configuration varies as a function of time, marks out in space the probability of finding matter or energy, and that gives it a material or energy aspect.

In the first half of the twentieth century, developments in quantum mechanics were extended. Remarkably precise results were obtained for the electromagnetic interaction using quantum electrodynamics. Following this, quantum chromodynamics explained the strong interaction, and this in turn led in 1984 to the discovery of the intermediate vector boson, and the electroweak interaction, the unification of the electromagnetic and weak interactions.

Heisenberg's uncertainty relations lead to wave-particle duality. By combining series of such waves it is possible, in principle, to construct any desired waveform. If we think again about the analogy of waves on the surface of a lake, it is possible to excite trains of waves in such a way as to create all sorts of interference patterns: a dip here, a flat surface there, and even quite large peaks in other places. A large peak would be the most striking feature on the lake surface; equally, the wave-particle picture tells us that the wave peak is the most likely place to find the particle. The narrower the peak, the higher the precision with which the particle's position is known. In order to make a very narrow peak through interference of waves, it is necessary to use wave trains in which the shortest wavelengths are at least as small as the width of the desired peak; with long waves we can only make figures with indistinct outlines. In summary: to localise a particle well with an uncertainty X, we must use waves with length shorter than X. Recalling de Broglie's relation, this length is Planck's constant h divided by the momentum of the particle; reversing the statement we can say the momentum is given by h divided by X. Another way of putting this is to say

uncertainty of position times *uncertainty of momentum* $= h$

which is Heisenberg's relation. The relationship for uncertainty in time is derived in a similar fashion.

The wave-particle phenomena have important consequences in experimental physics. One way of determining the structure of matter is to bombard it with particles. The trajectories of the particles following such bombardment tell us about the structure of the matter under investigation. To detect fine detail in this structure it is essential to know the positions of the particles very precisely, and for that reason they have to have large momenta (to minimise the uncertainty in position) and therefore high energies. This is one of the essential motivations behind the construction of large particle accelerators, the other being that if the incoming particles have lots of energy, the structure can be modified and new particles created.

We can now also see why physicists are interested in the very early universe, right after the Big Bang. Particles had very high energy at that time, because of the very high temperature, and their behaviour tells us something of their structure.

Strange particles

Antiparticles

Another surprising entity, and the real surprise is that it can be described as an entity at all, is the vacuum. Common sense tells us that a vacuum is nothing, or more precisely that it is empty space. However, in general relativity theory the vacuum isn't simply that. There are model universes (that is, there exist solutions to Einstein's equations) in which there is no matter; for example, empty hyperbolic space or empty Euclidean space, both of them expanding.

For quantum physicists the vacuum is much more complicated. Heisenberg's uncertainty relation for energy means that the energy of an electron, or any particle, can fluctuate. A similar conclusion is valid for the vacuum itself. Throughout the vacuum energy can appear as a result of a quantum fluctuation and then vanish into nothing, always provided that, in accordance with the uncertainty relation, the bigger the fluctuation, the shorter the time for which it will last.

But what are the consequences of this energy? At this point another astonishing quantum property of the vacuum comes into play. The theorist P. A. M. Dirac predicted this property in the early years of quantum theory, and his views were brilliantly confirmed shortly afterwards with the discovery of the positron, or positive electron, which is similar in its properties to the familiar electron carrying negative charge.

It is hard to understand the surprising feature discovered by Dirac; it is best to consider it in easy stages. Our first approach is to take an analogy that is so simple that it seems absurd. Imagine a row of books on a library shelf. If you remove a book, you can consider that you have created two 'objects': the free book (in your hand) and the hole in the row. These two objects can then be recombined by the simple action of replacing the book on the shelf; that's all there is to it. Although this appears very simplistic, it introduces the principle of creating and annihilating a particle and its antiparticle.

But, is the hole on the shelf really an object, just like the book that has been removed? It certainly has some properties in common. For example, just like a real book, the hole can be moved along the shelf, by the simple action of sliding neighbouring books one by one along the shelf. And the work needed to move a hole is the same as that needed to move a real book. So the hole behaves as if it had the mass of an actual book.

We can push this picture further. Suppose each book has an electrical charge, say -e like an electron, then the hole will behave as if it carries the opposite charge +e. To demonstrate this all that is needed is to connect the terminals of a battery to either end of the shelf, with the positive terminal on the left and the negative terminal on the right. Then the negatively-charged book immediately to the right of the hole will slide left towards the positive pole, causing the hole to move to the right; then the process repeats with the next book. In this manner the hole travels to the right; in so doing it behaves as if it carries a positive electrical charge +e.

Clearly it is a big jump from a book to an elementary particle. Nevertheless, this analogy recalls Dirac's original formulation. So far as he was concerned, the vacuum (the library shelf) is filled in a very compact fashion by electrons (the books). Within this environment a fluctuation occurs with enough energy to make an electron pop out (like removing one book), and thus a pair is created: particle and antiparticle, that is electron and positron (the book and the hole in the shelf).

The hardest part of this theory to take on board is the notion that the vacuum, which generally appears empty, is somehow packed out with electrons. But in making these statements we do need to choose our words carefully because experiments really have shown that the vacuum can produce particle and antiparticle pairs. To turn our analogy on its head, that's a bit like saying the only thing that happens in a library (the vacuum) is the creation of book and antibook pairs. Clearly such a vacuum is not the entity we normally associate with that word!

Actually, it is not necessary to wait for chance to cause fluctuations in order to obtain pairs of particles and antiparticles. They can be made by means of conventional energy and without any time limitation. For example, a particle can be bombarded with another that has been given very high energy in a particle accelerator, and the collision releases enough energy to make such a pair. All that is required is that the energy released is greater than the mass energy of the pair, that is, it must be $2mc^2$. In such collisions it is possible to create pairs of protons and antiprotons (with charge $-e$), pairs of electrons and positrons, and so on. By combining a positron and an antiproton, an antiatom of antihydrogen can be made. In principle, all kinds of antimatter can be fabricated: antiatoms, antimolecules, antimatter in general. Energy is thus transformed into matter in association with an equal amount of antimatter, in strict accordance with the relativistic law of the equivalence of mass and energy. And all this is possible because of the quantum properties of the vacuum.

One of the characteristic properties of antimatter is its tendency to be annihilated by the corresponding amount of matter any time they should meet, with a consequential release of mass–energy. However, antiatoms of antihydrogen can be confined by containers with magnetic walls, rather than material walls, and thus it can be stored readily. Such antimatter, so long as it is isolated from ordinary matter, is just as stable as ordinary matter.

Virtual particles

Newton admitted that gravity acted instantaneously through action at a distance. So one mass acts on another by means of a force whose value is given by Newton's law, calculated using the positions of the particles at that instant. This approach to a theory of gravity rests upon a concept of absolute time.

With the advent of special relativity, however, this notion of absolute time was wrecked. Henceforth lengths and time intervals were to depend on the motion of observers with respect to the observed phenomena, and the absolute would become something more general, the spacetime interval. This can be thought of as the hypotenuse of a right angled triangle in which one side is a conventional observed length and the other a time interval measured between two 'events' in spacetime.

The remarkable independent character of the spacetime interval is the foundation of special relativity theory. In the popular mind relativity might look like a retrograde step since it involves losing the conservation of length and time. In fact it is nothing of the sort. The conservation of intervals within spacetime is a more general statement; this powerful generalisation gives a stamp of authority to spacetime itself.

Electromagnetism falls readily into place in this framework. The concept that electromagnetic interactions are due to a field represented by a tensor in spacetime emerges naturally in the theory. The tensor is a kind of force, in four dimensions, which never suffers the drawbacks of the concept of force in the Newtonian theory. In the relativistic picture the electromagnetic interaction between two particles operates through the four dimensional 'superforce' and its field.

With advent of quantum mechanics, the dual wave-particle properties of light were expressed through the quantisation of the electromagnetic field. This operates through interactions between quanta, in this case packets of light or photons which behave like particles. These ideas were developed between about 1920 and 1950 from Dirac to

Feynman, and they resulted in unprecedented precision in calculations which involved photons.

It is thus that in relativistic and quantum physics the viewpoint is completely at odds with the Newtonian picture. The interaction between two particles is now described in terms of an exchange between them involving one or more other particles which serve as intermediaries. For example, an electron and a proton interact through an exchange of photons. These photons are the 'vector', or carrier, particles of an electromagnetic interaction in which the charged particles, the electron and proton, are the actors.

Let us now reflect on how all this came about. We represent the spacetime of relativity by a stack of transparent sheets on a table, with the most recent times on the top. An electron and a proton have world lines in the vertical direction (figure 13). At a certain instant of time the electron emits a photon. In reaction this jolts the world line of the electron and at the same time nudges the world line of the photon to the right. If, subsequently, the photon world line meets a proton world line, the photon is absorbed at the meeting point, and its world line is deflected too. The net result of such processes is that an electron and proton travelling freely and independently in spacetime find their paths distorted. This is a valid way to imagine the interaction between two charged particles.

In fact, for this to occur, the proton could also first emit a photon. Each particle could also emit one, acting in concert. Or the electron emits a photon and then absorbs it again; likewise the proton; and this could happen many times... Seemingly there is an infinity of possible interactions, and that embarrasses theorists because the result ought to be an interaction of infinite intensity! However, by a mathematical process whose logic leaves a little to be desired, it is possible to calculate results with extraordinary precision. That alone justifies the theory, at least for now.

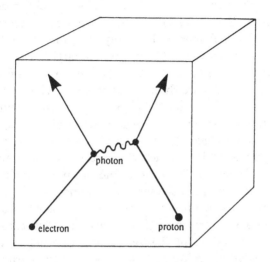

Figure 13 The interaction of an electron and a proton through photon exchange.

We shall now look in more detail at the emission of vector particles by real particles, and the energy which is required for their creation. Heisenberg's uncertainty relations tell us that this energy is available for a limited amount of time. During this limited time, the vector particle can travel as fast as light itself and go a certain maximum distance, called the Compton length. In making the calculations, the Compton length is given by h/mc.

From this expression it is clear that the virtual particle travels a distance inversely proportional to its mass. To carry out its role as a messenger (or carrier of a force) it must reach its destination in such a way as to allow the energy to be conserved. At the destination, the real particle will receive the virtual particle. During its journey, which can only happen at all through the protection of the Heisenberg uncertainty, this vector particle behaves in fact as a virtual (that is, non-real) particle.

For the electromagnetic case the mass of the photon is zero and so the virtual vector particle has an infinite domain. In principle it can travel to remote destinations. According to hypotheses that look more or less okay, the same would hold for the graviton, which would be the virtual particle carrying the gravitational force. For the nuclear interaction, in its form that we have known for more than 30 years, the carriers, or virtual particles, were pi mesons (or pions). The most recent ideas about nucleons are that they are assemblies of quarks; quark-quark interactions take place via gluons, which also have zero mass. For the weak interaction the vector particles have huge masses, one hundred times more than ordinary nucleons. In consequence the range of the force is extremely small: 10^{-15} cm.

The above is a brief summary of the basic ideas of how real particles interact through the medium of virtual vector particles. The masses of the vector particles determine the ranges of the different interactions. They run from gravitation, which encompasses the whole universe, down to the weak interaction which embraces no more than one-hundredth of a nucleon.

Quarks

We have only tackled the question of the strong (or nuclear) interaction in the last thirty years or so, during which time we have thought that it works when real particles, like protons and neutrons, exchange virtual particles, such as pi mesons or pions. In recent years the number of elementary particles discovered in experiments using accelerators has increased enormously, and now comfortably exceeds a hundred. This situation has parallels with that of the hundred or so atomic nuclei, the theory of which was simplified at a stroke, namely by combining just two particles: protons and neutrons.

In 1964, Murray Gell-Mann proposed a further simplification: protons, neutrons and the hundred or so elementary particles resulted from combining a few even more elementary particles: quarks. To keep things simple, but with no loss of basic concepts, we shall limit ourselves to the particles we met in our first tour of the universe: protons and neutrons, or nucleons, and their vector particles, the pions.

Gell-Mann considers these as assemblies of two types of quarks: u or up quarks and d or down quarks; the u quark has charge $2e/3$ and the d quark $-e/3$. Here we have introduced the revolutionary concept of one-third charges for elementary particles, which physicists had previously considered indivisible.

Furthermore, quarks are endowed with a new sort of charge or quantum number, in addition to their fractional electrical charge, and this is somewhat fancifully called 'colour'. This colour charge is not directly related in any way with the electrical charge of electromagnetism or the weak charge associated with the weak interaction. Nevertheless, like both of these latter charges, the colour quantum number determines, through complex equations, the strength of the new interaction associated with the quarks. The colour charge can assume one of three values, R, G or B, for each quark, representing red, green, or blue. Finally, for every quark there is an antiquark with the opposite properties associated with antimatter.

In this theory, nucleons are systems made of three quarks, and their vector particles are made of a quark and an antiquark. For example, the proton is uud and the neutron udd; the neutral pion is u and anti-u, and the positive pion is u, anti-d.

The properties of the nuclear interaction, specifically, the strong interaction, can then be derived from the theory governing the interactions between quarks. The theory of the strong interaction, which is called quantum chromodynamics, has been known for twenty years or so to be as reliable as quantum electrodynamics. This theory supposes that the interactions between real quarks take place through virtual (or vector) particles called gluons; these, like photons, have zero mass. They travel with the speed of light and the range of the strong interaction is, consequently, infinite, just as in the case of electromagnetism. None the less, nucleons themselves interact over a rather small range, 10^{-13} cm; we now have to consider them as complex systems. Their behaviour is similar to that of molecules; these also interact only over a short distance even though the electromagnetic force has an infinite range.

Unlike photons, which do not have an electric charge, gluons have the colour charge, represented by suitable combinations of the colours R, G or B. For example, an R quark emitting a R-G gluon becomes a G quark. This is a complicated effect, and it illustrates the fact that chromodynamics is more complex than electrodynamics.

There is another difference: although the active particles (electrons and so forth) and their vectors (photons) are observable in their free state, they are never seen as free particles in chromodynamics. Neither the quark nor the gluon exist as isolated free particles. This might seem to be a fatal blow to the theory, but theorists are resourceful people. One possible way out comes from the curious phenomenon of the polarisation of the vacuum. The fluctuations allowed by Heisenberg's uncertainty principle cause the spontaneous appearance of virtual photons and electron-positron pairs, and in the vicinity of an electron E there is a tendency for polarisation: the virtual electron is repelled and the positron attracted by the electron E. Around E there will be a positive zone and this partially screens the negative charge of E. Then, as observed from some distance, the apparent charge of E is weaker than its intrinsic charge. This vacuum polarisation has the effect that the intensity of the electromagnetic interaction, determined by charge, is stronger nearer the electron than it is far away. In the limit, in quantum electrodynamics, the naked charge of the electron is infinite.

The vacuum polarisation also comes into play for one quark, Q. It is surrounded by a sea of virtual gluons as well as pairs of quarks and antiquarks. These latter tend to mask the colour charge of Q. Gluons, unlike photons which have no electric charge, do have a colour charge. Rather than shielding Q, these gluons reinforce and extend Q's charge for rather complex reasons. The net effect of the vacuum polarisation is to amplify the intensity of the strong interaction at a distance and to decrease it locally. At very short distances the intensity asymptotically approaches zero. Through disappearance of chromodynamic forces or 'asymptotic freedom' quarks become independent of each other at very close distances.

Conversely, to separate them from each other we have to struggle against an interaction which becomes stronger and stronger, and thereby endows the quarks with more and more energy. A stage is then reached at which this energy, rather than being used to separate the quarks, instead gets squandered in the creation of pairs of quarks and antiquarks, which in their turn form new nucleons and pions. The net result is that quarks are permanently confined within heavier particles as a result of the fundamental nature of the chromodynamic force. This phenomenon of quark confinement, which arises via the polarisation of the vacuum, means that it is impossible to observe free quarks. Steven Weinberg, in his book *The First Three Minutes*, made a vivid analogy: it is as impossible as localising, or isolating, only the very end of a piece of string, which will break into two pieces with their own ends if you pull too hard.

This presentation of quantum chromodynamics moves us a stage beyond nucleons and the zoo of a hundred or so elementary particles discovered in the last quarter century. The zoo can be accounted for readily by means of just three families of quarks: the first with u and d, the second with s and c, and the third with b and t . Experimental evidence for the existence of five of these six varieties of quarks, or 'flavours', has been obtained in the last few years, and the existence of the t quark has been inferred at CERN. These experiments have also yielded values for the quark mass, which turns out to be some tens of times larger than the proton mass.

As a matter of fact, the second and third families of quarks only manifest themselves under quite exceptional circumstances such as very high energy collisions between particles and antiparticles. These conditions are not encountered in ordinary matter, where the first family of u and d quarks suffices.

This triplet of families is repeated in another class of heavier particles, the leptons, examples of which are the electron and the neutrino. Two other, much more massive, 'electrons' have been discovered in cosmic rays; they are the muon and the tau meson, which have, respectively, 200 and 3500 times the mass of the electron. With the former we can associate a muonic neutrino, just like the neutrino associated with ordinary electrons in beta decay, which we call an electron neutrino. There is every likelihood that a neutrino also exists that corresponds with the tau meson.

From the above we see that the nuclear interaction, as expressed through the strong interaction, is on an equal footing, from the standpoint of the theoretical formalism, with the electromagnetic and weak interactions. All three use vector particles and a similar fundamental picture, or metalanguage, to model what is going on. The similarity of description is the starting point for tentatively suggesting that the four forces in the universe could be unified into a single scheme, and from that we would learn new things about the fundamental nature of matter.

Elementary particles

We have now reached the stage where we can give a table of the elementary particles. The first major subdivision is between actor particles and vector, particles. The actor particles are those which, in one way or another, constitute the basic building material of the tangible universe. Within this constituency there are three families, and only one of them plays an important role in ordinary matter, by which we mean the stuff we see in the cosmos today. The other two families were at the fore in the first moments of the Big Bang, when very high energy phenomena were the norm, and they are also encountered in cosmic ray collisions.

The first family can itself be split into two sections: quarks and leptons, or in the round, heavy particles and light particles. The quarks comprise the u and d, and the leptons are the electron and the electron neutrino. So far as electrical charge is concerned, this first family has the values in the following table.

E	n	u	d
e	0	$2e/3$	$-e/3$

Each of these has its antiparticle, with opposite charge. So far as colour is concerned, leptons do not feature, but the quarks have the colour charge R, G or B. Lastly, from the standpoint of the weak charge, only left-handed particles and their right-handed antiparticles feature with charge +1/2 or -1/2 (expressed in arbitrary units of weak charge).

The other two families are related to the first in the following way:

	leptons	quarks
second	muon, muon neutrino	c s
third	tau, tau neutrino?	t? b

The charges associated with the three fundamental interactions are the same for all three families but the masses in the case of the second and third families are much higher. All of the vector particles have spin 1/2 and they are therefore fermions.

In the case of the electromagnetic interaction the vector particles are photons, which have no charge of any kind and no mass. The strong interaction takes place using gluons, which have electric charge in units of $e/3$ and colour charge corresponding to differences between the primary colours R, G and B of the quarks: G-B, B-R, and R-G. Lastly, the weak interaction proceeds through the W and Z particles, which have charge 0, or $\pm e$.

All the vector particles, which are the servants of the interactions, are actually quanta of the corresponding force fields; they have spin 1 and are therefore bosons. Just for the record, it is worth noting that the gravitational field would be carried through gravitons with spin 2, but no mass or charge.

The interactions of nature

A table of interactions

Before we embark on the larger question of how to unify the
interactions of nature, it is sensible to summarise each of them
in a table giving their main properties. The mathematical
jargon is pretty complicated, but these tables correspond to
matrices describing transformations; through group theory,
these matrices give a complete account of the various states of
the particles and the interactions between them. Although the
tables seem simple, they retain a trace of the underlying
physical processes.

To keep things simple we will concentrate on fermions
in the first family, of which there are eight. They are: the
electron, the neutrino, and six quarks, u and d each in three
colours. Of course, there are corresponding antiparticles which
we shall ignore for the present.

The simplest interaction of all is electromagnetism.
This is an engagement between particles that have electrical
charge, such as a pair of electrons, which we will call E. The
vector boson is a photon, p. In the table a row represents a
fermion that emits a boson and a column represents the
fermion that receives it. So, there is only one way of putting the
photon in the table:

	E
E	p

The weak interaction is found among particles that
carry the weak charge. It can be boiled down to weak
interactions between the pair left-spin electron e_l and left-spin
neutrino n_l, or the quark pair left up, u_l, and left down d_l. The
vector bosons in this case are the famous W particles, and we
need two tables:

	e_l	n_l		u_l	d_l
e_l	W0	W-	u_l	W0	W+
n_l	W+	W0	d_l	W-	W0

From this table we can see that a quark u which carries a charge $+2e/3$ can turn into a quark d with charge $-e/3$ by emitting a W+ boson; electric charge is conserved in this process. Armed with these tables we can construct the appropriate layers in a spacetime diagram showing the natural disintegration of a proton into a neutron (figure 14).

At the base of this diagram the neutron is represented by the three world lines of the three quarks *udd* travelling as a group. At some instant one of the *d* quarks emits a W-, and thus becomes a *u* quark. After this event the three quarks *uud* form a proton. Meanwhile the W- disintegrates into an electron and an antineutrino. The net result of this process is that the neutron has turned into a proton, an electron, and an antineutrino.

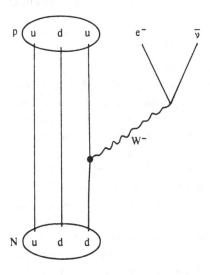

Figure 14 Disintegration of a neutron into a proton, an electron and an antineutrino through the weak interaction.

Finally, the strong interaction is uniquely given by the colour charges between quarks, the vector bosons in this case being gluons, g. The table has three rows and three columns. For *u* quarks for example it looks like this:

	uR	*uG*	*uB*
uR	g(R-R)	g(R-G)	g(R-B)
uG	g(G-R)	g(G-G)	g(G-B)
uB	g(B-R)	g(B-G)	g(B-B)

so a *B* quark emitting a (B-R) gluon becomes an *R* quark.

The fourth interaction is gravitation, and the theory for that involves gravitons of spin 2 as the vector particles. This theory is not so well developed as the others, and right now it is being attacked from every angle by numerous theorists who hope to unify all these interactions into a single theoretical framework.

Unification of the interactions

The first unified theory, of two interactions, was accomplished in 1967 by Weinberg, Salam and Ward, who showed how to represent the electromagnetic and weak interactions through a new concept, the electroweak interaction. The only price paid for the successful unification was the introduction of yet another particle, the Higgs boson (named for Peter Higgs of the University of Edinburgh), which is massive but has no spin. This particle enters the theory in a bizarre fashion.

When the vacuum is in a minimum energy configuration it contains a certain quantity of energy which is measured by the Higgs parameter. To a certain extent this parameter resembles the intensity of luminous radiation where the intensity fixes the number of photons. In order that the vacuum be deprived of Higgs bosons, it has to be provided with enough energy to quench their intensity. Conversely, this means that the 'truly' empty vacuum, without any Higgs bosons, is rich with latent energy wanting to makes its presence felt. What we are seeing here is another amazing property of the physical vacuum in advanced quantum physics. This dynamic aspect of the vacuum, which we shall dub 'vacuum dynamite', plays a crucially important role in the theory of the inflationary universe: by injecting an enormous amount of energy into the nascent universe, it blasts it on its way in a manner far surpassing the extraordinary violence of the initial stage of the classical Big Bang.

A further property of Higgs particles is to increase the mass of the W particle associated with the weak interaction. Furthermore, the W_0 boson thus modified is able to interact with a photon and turn into another boson, the Z_0, which has a different mass to the W+ and W-. The electroweak interaction table linking the left-spin electron and left-spin neutrino looks like this:

	e_l	n_l
e_l	p; Z_0	W -
n_l	W +	Z_0

The all important confirmation of the electroweak theory came with the discovery in 1983 at CERN of the W+, W- and Z_0 bosons, and the measurement of their masses. These masses tied in exactly with the theory, and they are large: 90 times the proton mass for the W bosons and 100 times for the Z_0 bosons. With masses like these they earn the name intermediate bosons (there are thought to be yet heavier ones, as we shall see). And the discovery of the Higgs boson itself is something for the future...

The next stage of unification, called Grand Unification, is still highly speculative and all we can do for now is to sketch out a theory, which is yet to be confirmed or refuted. Let's combine two of the tables we have looked at: the one immediately above for the electroweak interaction, and the table with three rows and three columns for the strong interaction of quarks. If we align the diagonals of these two tables we get a table with five rows and five columns which is partially filled:

	e	n	uR	uG	uB
e	p;Z₀	W -		X	
n	W +	Z₀			
uR			g(R-R)	g(R-G)	g(R-B)
uG		X	g(G-R)	g(G-G)	g(G-B)
uB			g(B-R)	g(B-G)	g(B-B);

The top left section is filled with intermediate bosons and the photon. At bottom right we have our table of gluons associated with the strong interaction. To these we should add the intermediate bosons and photons that are also coupled to the quarks. Two sections are empty and these are marked X. The principle of Grand Unification Theory (or GUT) supposes that there are new bosons to fill the gaps. These vector particles carry forces that allow transformations from leptons (electrons and neutrinos) into quarks and vice versa.

The 12 new particles (12 is the number that emerges in the simplest case) carry electric charge, weak charge, and colour, and they allow interactions between leptons and quarks. As an example, which does not involve too many details of a very complex theory, the electric charge of the X particles has values $-4e/3$, $-e/3$, $+e/3$, or $4e/3$.

The mathematical combination in pairs of the five basic fermions of the table leads to the fermions with opposite spin and to antifermions. If we do this we get around 30 new X bosons with various charges. All of them correspond to the first family of vector particles associated with the electron. Now repeat this operation for the second family, which ties in with the muon, and the third family, associated with the tau meson. In this way we get a table for all of the elementary particles in their various states, which follows a highly symmetric and rather simple scheme, as typified by the last table.

What masses do these hypothetical X bosons have? They turn out to be extraordinarily huge, some 10^{15} times the mass of the proton. This is about the same as the mass of a living cell 10 microns in diameter. Yet the distance over which they exert an appreciable force is fantastically small, and it is given by the Compton length, 10^{-29} cm.

Never before has physical theory landed us with particles this tiny. The size boundary was marked at 10^{-15} cm by intermediate bosons a hundred times heavier than the proton. When one thinks of the enormous efforts made at CERN, in which gigantic accelerators were used to detect the W and Z bosons, it is clear that the far more massive X bosons are out of reach of present techniques. With the exception of a few major predictions, such as the exact numerical equality of the opposite electrical charges of the proton and the electron, the proof or demise of Grand Unification Theory will have to take place in another domain. One such arena is the Big Bang itself where the available energy was large enough to create X particles and these should have left observable traces, such as evidence for the inflationary universe. A different circumstance where very rare effects predicted by the theory might be detectable is in very large quantities of matter observed over a long period of time; here one is trying to observe the disintegration of the proton.

There is a very large disparity between the masses of the vector particles in the electromagnetic (photon), electroweak (W, Z boson), and Grand Unification (X boson) theories: respectively they are 0, 100 and 10^{15} times the mass of the proton. This has immediate consequences. The first is that although these interactions have all the appearances of being symmetrical and unified, they do not manifest themselves on an equal basis under all circumstances. In particular, at low energies, or low temperatures, or over large distances, the parts of the theory that use light vector particles take over. But at higher energies, or when temperature is increased or on very short distance scales, the other particles come into play.

Just this situation existed in the very early universe, with its extremely high temperature. There the interactions were unified and acted in perfect symmetry. Then, through cosmic time, the temperature fell and this symmetry was progressively destroyed. At 10^{27} K the X bosons could no longer be created spontaneously from the ambient energy; then at 10^{15} K intermediate bosons were toppled also. In cosmic time these critical temperatures correspond to 10^{-35} second and 10^{-11} second; these are pretty tiny compared to the epoch 1 second after the Big Bang when the universe as we perceive it now came into being.

The events of those times are particularly interesting. As the temperature fell, so the symmetry of Grand Unification was broken in two critical stages. This led to a decomposition from which sprang the three interactions that we see in the natural world today: electromagnetism, the weak interaction and the strong interaction. (We cannot account for the gravitational interaction and that is left on one side for the present.) This spontaneous symmetry breaking plays a role of the highest importance in releasing the dynamite of the vacuum as implied by the Higgs particles.

The disintegration of matter

Twenty five years ago astronomers discovered the microwave background radiation. This breakthrough allowed us to pose the big question, once we had traced the history of the universe from the first second down to the present time in tbroad terms: what is the ultimate fate of the universe? There are two possible tracks: indefinite expansion, or a new contraction. Which route will the universe choose? General relativity has guided our thoughts so far. We made a long journey of exploration, using quantum physics; now we find looming on the horizon a concept that is incredibly overwhelming: we can paint a scenario in which the matter of the very universe itself simply disintegrates! According to this picture, the most fundamental material particles will eventually fade away.

We are not talking here of the familiar disintegration that turns neutrons into protons, or the process that converts uranium into lead. In those processes matter (nucleons) continues as matter but in a different arrangement (either different nucleons or different combinations of them). Nor is it the sort of destruction that takes place when a particle and its antiparticle (matter and antimatter) combine and annihilate, with a consequential release of the energy that had previously helped to create the antiparticle. What we are now about to encounter is the dazzling concept of matter going straight to energy, that is nucleons becoming photons. And so the awesome possibility of the complete disappearance of matter in the long run raises its head.

Metaphysically this phenomenon is crucial to the ultimate fate of the universe. Relativistic models of the universe have already surprised us with scarcely reassuring scenarios on timescales of hundreds of billions of years: they tell us that we can choose between the final implosion of the fireball or the absolute cold of never-ending expansion. And now we have something even worse to take on board! If the future of the universe is along the cold track, not only do the energy resources get spread ever thinner, but the basic building blocks, the flesh of the cosmos if you will, the nucleons, disintegrate into pure photons of electromagnetic energy.

Evidently this is not going to happen tomorrow, nor even in tens or hundreds of billions of years. This particular eventuality is more than 10^{32} years into the far future. Anyhow in less than 10^2 years the destiny of everyone now living will have been determined!

Let us turn back to the transparent layered model of spacetime, and consider the most common constituent of the universe, the hydrogen atom (figure 15). Its world line, drawn vertically upwards, comprises two such lines:

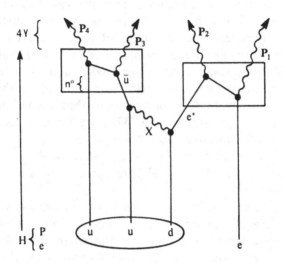

Figure 15 Disintegration of a hydrogen atom to four photons through the Grand Unification interaction

There is one for the proton and another for the electron, travelling in parallel but spaced laterally by 10^{-8} cm. The worldline of the proton is in turn made up of three component lines representing the quarks u, u, and d, also travelling as a tightly bunched group, in this case spaced some 10^{-13} cm from each other.

The table for Grand Unification tells us that a quark d with electrical charge $-e/3$ can be transmuted to a positron E+ by emitting an X boson with charge $-e/3$. For its part, the electron within the hydrogen atom can emit a photon p1 causing a kink in its world line. At this stage it can meet up with the positron and their two world lines join at a further kink, whence the photon p2 travels. The right hand portion of this Feynman diagram shows the typical annihilation of an electron and a positron to give two photons p1 and p2. The total effect is that the electron in the hydrogen atom and the positron created when the quark in the proton emitted an X boson have

X boson have been annihilated into two photons. They are gamma ray photons in fact, on account of the masses involved.

As for the X boson, in all probability it gets recaptured by one of the proton's u quarks, with charge $2e/3$, which promptly turns into an anti-u quark with charge $-2e/3$.

What remains of the proton? Quite simply, a u and an anti-u quark, that is, a neutral pion, which itself dis-integrates like electrons and antielectrons to produce two further photons p3 and p4 (these are still more energetic gamma ray photons).

The net result of all this is that a good, simple, hydrogen atom has 'vaporised' into four gamma-ray photons. This is the fate Grand Unification Theory projects for ordinary matter, so now we have not only replaced the illusory durability and solidity of matter, as symbolised by prehistoric stone tools, by a void sparsely populated by nucleons; we have gone further and said that the nucleons themselves might vanish into a puff of pure energy.

What is the probability that a hydrogen atom could be snuffed out in this peremptory fashion? In the process outlined above several conditions must be fulfilled. The strongest constraint is that the X boson emitted by the d quark in the proton must meet one of its two u quarks. However, Grand Unification Theory endows the X boson with an enormous mass; the virtual X boson in this interaction has a tiny range of 10^{-29} cms and it has to bang into the u quark within this distance.

This meeting takes place in an immense desert, or so it seems. The virtual X boson cannot go more than 10^{-29} cm on the average but, at the same time, the u quark is likely to be 10^{-13} cm (the diameter of the proton) on average, or 10^{16} times further away than the range of the X boson. In terms of relative size this is a bit like asking an organism the size of a virus to meet up with one of only two other viruses to be found somewhere in a volume of space as large as the Sun! These numbers, and the virus analogy, give us some insight into the incredible vacuum that exists inside a nucleon.

Before quantum physics had unravelled all of these secrets it seemed that a nucleon with a diameter of 10^{-13} cm had the impressive density of 10^{15} g/cm^3. What we have now revealed is that within the volume of the nucleon the space is as sparsely occupied as a sphere as big as the Sun with three viruses inside. From this fragile stuff we have to construct the entire universe!

Effectively, the enormous amount of space inside the proton means that the probability of decay is very small indeed. Present calculations give a lifetime of 10^{31}–10^{33} years. These estimates are truly enormous: more than one thousand billion billion times the actual age of the universe. However, physicists are devising means of measuring this huge lifetime.

The theoretical interest in this subject is so strong that it has stimulated experiments across the globe in attempts to detect proton decay. A moment's thought should suggest that this is not a simple matter of watching one atom for 10^{31}–10^{33} years in order to catch the moment at which it disappears. Instead the principle is to observe a huge quantity of protons, or hydrogen atoms actually, and count how many decay in some convenient time interval such as a month or a year.

For several years experiments undertaken in the USA, USSR, Japan, Italy and France have only yielded negative results. Each experiment has involved the monitoring of several thousand tonnes of matter. The observations are made in deep mines as far as 8 km below the surface, where the detectors are shielded from spurious signals from natural cosmic radiation by the overlying layer of rock. The conclusion from the experiments is that the characteristic timescale (or half-life) for proton decay is 10^{32} years at least. We must continue to hope that a positive result to this crucial test of Grand Unification Theory is not too far away.

Subquarks, supersymmetry and super-gravity

In Grand Unification Theory leptons (electrons and neutrinos) are on an equal footing with quarks; collectively they make a single category of fermions with spin 1/2. Interactions between these fermions are effected by the exchange of bosons with spin 1, photons, intermediate bosons, gluons, and X bosons; we also need to add Higgs bosons, with spin 0, and possibly gravitons, spin 2.

This basic series encompasses more than two dozen particles. If we include their antiparticles, and then multiply by three to incorporate the second and third families, we easily score over 100 particles. This situation recalls the familiar series of the atomic nuclei of the 100 or so elements, which have nucleons as their basic subcomponents. In the case of atoms we found that a rather small number of such substructures could account for the numerous observed combinations of nuclei. The nucleons themselves only appeared to be elementary when discovered because the theoretical techniques used to describe them did not extend far enough.

In our search for more basic subcomponents we have to look on an extremely small scale. Effectively, the enormous precision attained in quantum electrodynamics and in high energy collisions involving electrons shows that in all respects the electron behaves as a point down to scales of order 10^{-16} cm. Nothing indicates a substructure with a scale size greater than this.

Many theorists have thought about the ultimate structure of matter. At present the tentative ideas with some chance of success are quite numerous. That in itself indicates that none is convincing. To give an idea of the possibilities I am going to sketch just one such notion. Harari has proposed the existence of just two sub-quarks, named T and V. The first of these has electric charge $e/3$ and can assume colour charges with values R, G or B. The second sub-quark has 0 electric charge and the opposite colour charges. For this pair of sub-quarks there are the anti-sub-quarks T̅ and V̅. And that is all.

In this scheme, leptons and quarks result from very simple rules of combining T, V, T̅ and V̅: put together three sub-quarks or three anti-sub-quarks, excluding all other possibilities. Then, VVV will be a neutrino, T̅T̅T̅ an electron, TTV a quark u and V̅V̅T̅ a quark d, etc. The truth of this can be verified by adding the appropriate charges. The most remarkable feature is that these elements and the rather simple rules associated with them lead to the complete series for the first family and they give a good explanation of the mysterious relations between colour and electric charge, such as the fact that leptons do not have colour.

The proton in such a model is made of nine sub-quarks arranged as three groups. The Feynman spacetime diagram readily shows that proton decay can be explained by the simple exchange of a V and a T sub-quark between two groups, followed by the regrouping of the six sub-quarks in two different groups to make a neutral pion and a positron (figure 16).

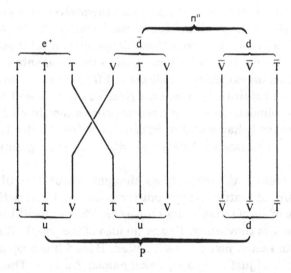

Figure 16 Disintegration of a proton into a positron and a neutral pion according to the sub-quark scheme.

These results are encouraging. At present a more sharply articulated theory is not in view and there are plenty of places to stumble on the road ahead. It is too early to say whether any of the hypotheses will have lasting value. How, then, are we to progress? It is now essential to probe the structure of the electron on scales smaller than 10^{-16} cm by using the most powerful particle accelerators available. Furthermore, we must observe proton decay unambiguously. Lastly, it would be helpful to have more information on the particles in the second and third families. Such endeavours could take a couple of decades. Despite the fantastic progress already made in uncovering the structure of matter, there is a long haul ahead before we will see the complete picture.

In the last few years, theoretical ideas directed towards a single concept of all-embracing symmetry have been floated. One of these theories is supersymmetry. This postulates that the fermions in the basic set of elementary particles (electrons, neutrinos and quarks) and the bosons associated with their interactions play similar roles, despite large differences in their statistical behaviour. In supersymmetry, spacetime possesses extra dimensions. These represent 'fermionic' coordinates in addition to the usual four dimensions. Spacetime becomes a many-dimensional superspace. Within this superspace the mathematical formalism possesses a symmetry that has aesthetic attractions for theorists. The theory naturally leads to a defined number of elementary particles, although that number is very large. The particles are arranged in pairs of fermions and bosons which correspond symmetrically. New hypothetical particles emerge from the scheme: a photino as the boson to partner the photon, leptinos, quarkinos, gravitinos, and so forth. There are many more besides these, but not a single one of them has any experimental support. Cosmologists, feeling perhaps a bit lost but nevertheless wanting to use such exotica to trigger the expansion of the universe, call these particles '...inos'.

Let's ask what new experimental techniques would test these hypotheses. Every scientific hypothesis must be confirmed experimentally, or else it is doomed to remain as wishful thinking or mere speculation. In this connection, it is regrettable that free-wheeling speculations have passed rather rapidly, without proof, into the everyday talk of popular science, when their basis is hardly stronger than a lot of pseudo-science. This will only lead to even greater confusion. However, supersymmetry has made one testable prediction: in the case of proton decay it suggests that rather than decaying to a positron and a neutral pion, the proton would become a muon and a kaon. Experimental tests of this are envisaged.

What becomes of the gravitational interaction in these efforts about unification? Within general relativity, gravitation takes its form and being from the geometrical curvature of spacetime. For some years now, however, a different approach has been considered. Following the success of quantum physics, theorists have explored the gravitational interaction as if it were carried by vector particles of spin 2; these are the hypothetical gravitons, which have never been detected. The two approaches — curved space and gravitons — seem diametrically opposed. Nevertheless, some ten years ago the proponents of supersymmetry showed that in one of its guises, supergravity, the superspace possesses geometrical supercurvature and supertorsion. In such a space, fermions of spin 3/2 (gravitinos) would interact with bosons of spin 2 (gravitons).

From an observational standpoint, in recent years the very detailed observations made of the binary pulsar have verified numerous phenomena predicted by general relativity. There just remains a tiny anomaly in the rate at which the pulsar is slowing, but this residue corresponds exactly to the energy that would be radiated as gravitational waves, again in accordance with general relativity. This virtually amounts to a proof for the existence of such waves, which is a way of saying that undulations in the curvature propagate like electromagnetic waves. If progress continues will we one day be able to detect the quanta of this radiation, the gravitons? And if so, why not the even more hypothetical gravitinos?

Despite reports to the contrary, we cannot yet see the light at the end of the tunnel on our long journey to a complete description of the nature of matter. The two golden achievements of twentieth century physics, the relativity and quantum theories, must be fused together in a convincing fashion before we can better understand the universe and its future.

6

The inflationary universe

Chapters 4 and 5 tackled the relativistic and quantum aspects of the universe, in accordance with the two major theories of the age, general relativity and quantum theory. Even though these have enjoyed much success in explaining numerous experiments, as well as observational data, they both have shortcomings and need a more complete synthesis. Supersymmetry, supergravity and even Grand Unification are promising ways to approach this goal but they lack experimental support, being only at the stage of preliminary outlines. However, the cosmologist Dennis Sciama has made this encouraging remark: 'It is hard to imagine that everything is wrong or illusory. We are witnessing the beginning of a new and imaginative scenario for understanding the universe'.

In Chapter 4 we played the game of trying to understand the universe, and took some risks. It is fun to launch out on a promising track, avoiding the pitfalls for the unwary and sidestepping the dead ends, in order to see if the chosen route will open up new horizons or lead to an impasse. In any case, to accompany a scientific mind voyaging a slightly dangerous but rational course is an interesting pursuit.

Weighing up the Big Bang

The Big Bang scenario, which has already been discussed in this book, has plenty of positive features. It emerges naturally from a well understood theory based on relativistic model universes. It gives a full account of the recession of the galaxies, the cosmic microwave background, as well as the primordial abundance of hydrogen and helium, which was fixed in the first fifteen minutes of nucleosynthesis. The theory also enables us to obtain relations between the age of the universe and the ages of the oldest stars observed, the density of the universe as given by the distribution of galaxies, and the expansion velocity indicated by the Hubble parameter. The theory is entirely consistent with the observed isotropy and homogeneity of the universe when considered on the largest scale.

Over and above this list of positive attributes there are some parameters that are not at all well determined. For example, the deceleration parameter (the rate at which the expansion is slowing down or speeding up) has not yet been measured observationally with enough precision. For the curvature of space we are no better off since we cannot as yet even tell whether space is spherical or hyperbolic. Possibly that is in itself an indication that we are close to or at the intermediate configuration of Euclidean space. Finally there is the cosmological constant, a mathematical term slipped naturally into Einstein's equations which represents a cosmic repulsion over very large distances; its value is unknown, although it is certainly small.

What negative elements do we need to consider in order to work out the pros and cons of the Big Bang model? There are niggling worries:

1. It cannot explain the almost perfect isotropy of the 3°K microwave background radiation. The argument here is quite simple. No matter where we look, the radiation is identical in a given direction and the diametrically opposed direction. But the two opposing regions that send us this radiation are so remote from each other that they can never have been in direct contact in the course of the expansion, nor has any indirect contact taken place. Each region is beyond the

cosmological horizon of the other. We cannot therefore account for the fact that they are in identical states physically since they seemingly have never had any way of 'communicating' with each other. This is the problem posed by isotropy.

2. The actual curvature of space cannot be far from zero, despite the fact that we cannot tell what type of curvature is present. And this is the case after 15 billion years of cosmic evolution. Equations governing the evolution of model universes tell us that, for such a small curvature today, the curvature had to be fixed very close to zero, with enormous precision about one second into the Big Bang. Another way of expressing this is to say that from the outset space must have been quasi-Euclidean. The Big Bang model gives no explanation of this at all. This is the problem of the flatness of space.

3. When the amount of matter in all galaxies and clusters is totalled and compared to the number of photons in the universe, principally those in the 3°K microwave background, it turns out that there is one nucleon for every billion photons. Why should the ratio of nucleon number to photon number be as small as one billionth? The straightforward Big Bang theory is unable to explain this to us. In fact this rather anodyne question is firmly rooted in the whole issue of the origin of matter in the universe.

The inflation of the universe

The vacuum phase transition

The 'dynamite' of the vacuum provides energy in the inflationary universe, and a phase transition releases that energy. A very familiar phase transition occurs when liquid water becomes ice. In physics a phase transition is the transformation from one physical state to another. Water, for example, can exist in three physical forms: vapour, or the gaseous state, in which molecules are widely separated from each other; the liquid state, in which the molecules are more crowded together in a disorderly way; and, finally, the solid or crystalline state in which the molecules are densely packed and forced to occupy a regularly-spaced lattice of places.

Water, therefore, has three possible phase transition, and two of these are very familiar : gas to liquid, and liquid to solid, which at normal atmospheric pressure take place at 100°C and 0°C, respectively. Here we are interested in the liquid to solid transition because this gives the best analogy with the 'dynamite' of the vacuum. The liquid and solid states have two essential differences:

1. The liquid state is completely isotropic: in liquid water there is no preferred direction so the properties of water are the same in every direction. In the crystalline state, by contrast, the symmetry is broken. Along the lines of the crystal lattice the physical properties differ from those observed in other directions. For example optical refraction, mechanical strength and other properties all depend on which direction through the crystal you choose to make the measurements. What this means is that whenever there is a transition between the two states, for example, from liquid to solid, the symmetry is broken in the physical theory that describes the properties of water.

2. The crystalline state has less energy than the liquid state. The evidence for this comes from melting ice: we have to heat it. The principle is exploited in the ice chiller packs used in a picnic cool bag or Esky: the heat needed to melt the chill pack deprives the fresh food of the chance of becoming warm. Conversely, when we turn liquid water to ice we have to remove heat energy, using a refrigerator. The energy difference between these two states of water arises because when the molecules are fixed in a regular crystal lattice they interact with each other much more strongly (through the outer orbital electrons) than when they merely glide past each other in the disordered fluid state. The heat energy is termed latent heat by physicists; it is the heat needed to bring about a phase change alone, without any change in temperature.

When we consider the properties of the vacuum in Grand Unification Theory the situation is rather similar. The very special sort of vacuum is either densely or sparsely populated with Higgs bosons (those hypothetical particles which have never been detected). The energy density of the vacuum, or its energy levels, depend on the density of Higgs bosons. Essentially there are two distinct states, such as we explored with water.

1. A high energy state, which corresponds to liquid water, where the density of bosons is zero and the vacuum is symmetric.

2. A state of much lower energy, like ice, where there are significant numbers of Higgs bosons; these behave in a manner that depends on their properties, just as there are preferred directions in a crystal lattice. This energy state only exists below a certain critical temperature (theory gives 10^{27}°K), just as ice requires a temperature below 0°C.

What this means is that below a temperature of 10^{27}°K the vacuum can make a phase change from a perfectly symmetrical state with no bosons to a state of broken symmetry with some bosons at a much lower energy level. When the vacuum passes from its first state, in which the vacuum is symmetric, to the second, unsymmetrical, state, the amount of energy released is enormous, according to the theory, because there is a large energy difference between the two states. This transition energy is, therefore, the 'dynamite' which is released by the phase change.

The inflation of space

Our basic model (in the grandiose fresco) retraces the history of the universe from one second after the Big Bang to the present day (15 billion years later). The model also gives us some insight into possible future scenarios, stretching hundreds of billions of years into the future. Notwithstanding these successes, it cannot give us any information about the more remote past, from time zero to one second after the Big Bang. The inflationary universe is the exciting scenario which we are going to insert into this era of cosmological prehistory.

Let's go back right away to the moment 10^{-35} second after the Big Bang. In everyday language, or even the language of cosmologists, it is rather hard to take such a tiny time

time interval on board. If we imagine stretching one of our seconds of time into 15 billion years, then that 10^{-35} second converts to 10^{-18} second, which is still fantastically small; it is, for example, 10,000 times smaller than the smallest time interval that can be measured in a laboratory. In spite of this, 10^{-35} second assumes significance in the theory, and we can take it as a great triumph of human endeavour that we are thus able to get so close to the time zero.

At 10^{-35} second, the highly condensed and extremely hot universe expands and cools, just like General relativity and the equations of state for relativistic universe say it should during the expansion. At this golden moment, which is about to be lost for ever, the temperature falls to $10^{27°}$. This is a memorable occasion because at such a temperature the photons have the same energy as X bosons, in accordance with Grand Unification Theory. Henceforth, therefore, X bosons can only exist as virtual particles, whereas previously they had existed in thermal equilibrium in numbers comparable to other particles. This, indeed is the reason why symmetry is going to be broken because, as we saw above, in Grand Unification symmetry breaking occurs at 10^{-35} second.

The main consequence of this is that the vacuum, which was symmetric, gets turned into a vacuum with broken symmetry via the phase transition. In fact, this phase transition does not take place instantaneously. There is a 'supercooling' process, just as with the phase change from water to ice: if water is cooled very slowly and uniformly it will remain liquid for several degrees below zero, provided it is free of impurities and isn't subject to any vibrations. This is supercooled water, with a temperature below zero. But any small disturbance, such as shaking it a tiny bit, will immediately make it turn to solid ice, which suddenly releases all the latent heat in one go.

Something similar to this process may happen to the universe. There may be a delay as the vacuum turned from being symmetric to a state of broken symmetry. Then each cubic centimetre of space, although cooled below $10^{27°}$K by the expansion, retains the high energy associated with the symmetric vacuum. Although this 'dynamite' does not explode, since the phase transition is not yet started, it does fundamentally change the behaviour of the expansion by speeding it up tremendously compared to the 'routine' Big Bang expansion.

This acceleration is the secret of the inflationary universe. The concept and its theoretical development due to A. H. Guth (Massachusetts Institute of Technology) and A. D. Linde (Lebedev Institute, Moscow) is a nice example of the universality of science. However, the concept is very hard to grasp. What happens is this: each cubic centimetre contains an enormous amount of energy, which it has on account of the symmetry of the vacuum. This energy is the same for every cubic centimetre, and this remains so even though the scale length of space evolves with the expansion. Considered from the standpoint of the vacuum this is perfectly simple since a cubic centimetre of vacuum is a cubic centimetre of vacuum even if the space in which it is located is expanding.

But if we now look at this instead from the point of view of space and how it behaves things are rather different. As space expands more and more energy is introduced into the universe as the total volume available for the vacuum increases. As an analogy, imagine space behaving like a block of rubber: the more it is stretched (or expanded) the more energy it has because the elastic fibres store energy. The vacuum behaves like a gas with 'negative' pressure. This negative pressure introduces an extra term, the cosmological constant,.into Einstein's equations. It acts on the expansion like a cosmic repulsion.

This cosmic repulsion, which is always of the same value and is inexhaustible, has the effect of spurring on the expansion of the universe at an explosive rate, exponentially in fact, because its strength is constantly renewed, rather than draining away with such efforts. The result is that space triples its size every 10^{-34} second after the epoch 10^{-35} second. The theory then suggests that the inflation lasts until the epoch 10^{-32} second, when the super cooling stops. Since there are 100 units of 10^{-34} second to use up before the elapsed time is 10^{-32} second, the tripling process takes place around 100 times: $3 \times 3 \times 3... =$ 10^{50} times altogether. This is the inflation era of the expanding universe.

In simple language we can try to recap on this extraordinary period. The universe starts out very hot and extremely condensed; the vacuum state is the symmetric one demanded by Grand Unification Theory. The universe then goes through a period of incredible expansion, driven by the energy of the vacuum itself, and this inflation lasts until 10^{-32} second. The dimensions of the universe are vastly dilated through the bonus of vacuum energy released by every one of the new cubic centimetres of space which are set free (out of nothing)! The most amazing aspect of this supercooling scenario is that every time a new bit of space is created, it also is provisioned with its 'dynamite' to keep the expansion going (figure 17). This is because every part of space is trapped into a higher-energy state than the state in which symmetry is broken. During the instant that the expansion takes place, every new cubic centimetre of new vacuum contains the same energy as the existing vacuum. The universe truly gets something for nothing! If these incredible ideas hold up as knowledge advances, there will be plenty of food for thought for philosophers!

What happens next? The supercooling must be over because the temperature is much lower than 10^{27}°K, much more than it would have been according to the more moderate expansion rate given by the standard Big Bang models. There is a sudden transition from the situation in which the vacuum is symmetric to the phase with broken symmetry; this cross over instantaneously releases all the pent-up energy of the 'dynamite' from the symmetric phase. This is the second event.

The third and final event in the new inflationary universe then occurred. The energy which was suddenly released throughout space heated the universe up again to nearly 10^{27}°. Myriads of Higgs bosons appeared at the moment of symmetry breaking, and with their considerable energies they created all sorts of particles. For example, fermions such as nucleons and antinucleons, and bosons such as photons, W and Z bosons.

Space, which had been only a symmetric vacuum with symmetric forces, suddenly thronged with particles.

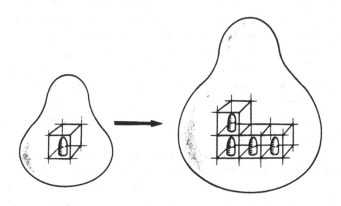

Figure 17. Any portion of spacetime contains more and more centimetre cubes as it progressive;ly expands, and each of them has its own share of dynamite, or vacuum energy.

At 10^{-32} second the universe really took on the traditional form that we recognise. The vacuum had lost its energy content and its place was taken by particles of every species. The environment had a temperature of around $10^{27°}$. The expansion then followed the more moderate rate of the standard Big Bang model which has lasted from one second after the beginning of the grandiose fresco, right down to the present day, 15 billion years later. If the ideas outlined above are correct, nobody should dare suggest that the first 10^{-32} seconds of the universe are negligible compared to the next few billion years.

The universe in an apple

To summarise the major consequences of the era of inflation we shall need to use the power of ten notation because the values we encounter are way beyond those humans can readily comprehend. The easiest way of describing the events is to take them chronologically. In fact the observable universe does stretch for nearly 15 billion light years, because the 3°K background radiation, the most distant 'object' that we can see, comes from close to the cosmological horizon. If we back track to 10^{-32} second, the moment when the inflationary era finished, we find that it is valid to use the standard Big Bang model and the equations of general relativity. These tell us that the size of the now observable universe was then no more than 10 centimetres across, about as big as a nice apple! And when the inflationary era started, at a mere 10^{-34} second after initialisation, this cosmic apple was 10^{50} times smaller, a mere blip 10^{-49} cm from end to end.

You might well ask what significance, physically, can such a tiny diameter possibly have. This is a crucial question, as a matter of fact. The answer requires us to consider the size of the cosmological horizon at that instant. That dimension is given by the maximum distance travelled by a signal going at the speed of light, which is 10^{-24} cm in 10^{-34} second.

At this point we need to return to supercooled water that turns into ice. In this sudden transition the water does not solidify into a single perfect crystal. Instead it makes numerous tiny crystals, chaotically jumbled together. This illustrates two things. Firstly, when the symmetry of liquid water is broken into the asymmetry of crystallised water, the symmetry breaking manifests itself in the infinite variety of ways in which the individual crystals can be arranged spatially. Secondly, during the sudden phase change to ice, no information is available to instruct the crystals to arrange themselves in any preferred manner. Each crystallisation takes place individually, so collectively they are disorderly. This results in a mass of independent crystals, the sizes of which depend directly on the speed at which freezing takes place. Rapid freezing means many tiny crystals.

The situation in the universe is rather similar. When the supercooled symmetric vacuum became the frozen vacuum of broken symmetry, an ensemble of compact domains resulted, each with its own properties. The scale of these 'crystals' with different vacuum properties is given by the cosmological horizon, 10^{-24} cm, the maximum distance for the coordination of messages. An immediate consequence of this is that the scale of our universe, 10^{-49} cm, was 10^{25} times smaller than the scale of the 'crystals' when symmetry was broken.

Now we have reached the first crucial result: the inflationary universe model implies that the observable universe was tiny compared to the 'crystal' to which it belonged. It occupied only a miniscule part, so tiny in fact that it could not fall between two stools and inherit bits of two 'crystals'. This guaranteed that it would have the same properties throughout. And from this came the isotropy and homogeneity of the cosmos we see today, a feature that cannot be explained by the standard Big Bang model.

The second major consequence is this: if our 'crystal' came in a part of space having a certain curvature, and it is entirely reasonable to think that this was so, then the radius of curvature would now be extremely large on account of the factor of 10^{50} in the inflation. In this way we can naturally explain why the curvature observed today is so small (radius very large). In the standard Big Bang it is impossible to account for the fact that space is practically Euclidean, which is the special case between spherical and hyperbolic space.

These twin achievements emerge automatically from the inflationary scenario and they are among its most attractive features, although they do not prove it. They are encouraging results, and justify our reflecting on the concepts, processes and astonishing implications uncovered by the theory. We have paid the price, and an inflated one at that, to learn about the immensity of the cosmos today, and the reasons for it. All of its material, its countless stars and its billions of galaxies seemingly without end, was once no bigger than an apple, at the end of the inflation era. And this apple came from a fragment of vacuum billions and billions of times smaller than the nucleus of an atom.

The second detonation

The origin of matter

At the end of the inflation era the universe emerged from the vacuum state described by Grand Unification Theory. It found itself well supplied with space and energy. That was the moment in space and time when matter was created from energy. The properties of the vacuum as described by Dirac give an example of how pairs of particles and antiparticles can come from pure energy. Two photons, for example, can create an electron and a positron in the vacuum so long as the two photons have more energy than the rest mass energy of the two particles. This is so if the temperature of the radiation that provided the photons in the first place is more than a critical energy, $10^{10\circ}$, in this case.

After the phase change, the universe was sig-nificantly reheated, and from this process a bountiful mix of particles and antiparticles resulted. Statistical equilibrium demanded that particles were present in equilibrium, so that for every electron or nucleon there was a photon or boson, respectively.

However, the resulting particles and antiparticles have the bad tendency to recombine and make photons as soon as the temperature falls below the critical value because the statistical equilibrium is broken. Matter and antimatter annihilated each other and became pure energy below the critical temperature. This seems to imply that neither matter nor antimatter should have survived in the universe. If such an outcome is in principle possible for a universe we simply have to remember that that was not the fate of our universe, since we are here.

How did this come about, and to think of it, how can we account for the observational fact that for every nucleon in the universe there are a billion photons? This would have happened if, rather than making one particle and one antiparticle, nature had arranged for 999,999,999 anti-particles for 1,000,000,000 particles; then, after annihilation, we would be left with one particle for every billion photons.

It was Andrei Sakharov who indicated, in 1967, what sorts of conditions would be required to cause this small asymmetry between the production of particles and antiparticles. A dozen years later it became clear that the conditions could be satisfied by Grand Unification Theory. This comes about because at one stage the transformation of antiquarks to antileptons, through X bosons, is slightly easier to achieve than the transformation of quarks to leptons. The calculations leading to this conclusion are, nevertheless, rather hesitant. This is so much so that at a symposium one of the theorists involved said that to get a number from this muddy yard is the same as trying to pluck a rabbit from a soup of amino acids! However, a one billionth survival advantage for particles over antiparticles is possible and this resolves a problem that is inexplicable in the Big Bang theory.

In summary, we can rather provocatively (or even outrageously) say that the reasons why matter exists in the universe today are: firstly because matter is intrinsically unstable. The small asymmetry between particles and antiparticles is in part due to those processes that also cause the proton to be unstable. Secondly, there is matter now because there was no matter at the beginning, contrary to ideas that have hitherto been accepted. And thirdly, we should add this: the curious state of apparent contradictions will not last forever, because protons should finally disintegrate. This then is the way we can now see the birth of the universe in accordance with grand unification and the inflationary universe picture.

The longest second

The wild course along the pathways to the universe engineered by theoretical physicists and cosmologists is now approaching its conclusion. The phase transition at 10^{-32} second suddenly causes reheating of the universe, and the resulting sudden burst can perhaps be called the second detonation of the universe, giving birth to the standard Big Bang.

What is the fate of the bosons? The X bosons created during the second detonation rapidly disappear because the temperature soon falls. Just before, they slightly favour a tiny excess (one billionth) of quarks over antiquarks.

The temperature falls as the universe expands and a point is reached at which the electroweak force also decomposes. It splits into the weak interaction and the electromagnetic interaction in a new symmetry-breaking event. This leads to another injection of energy into the universe at around 10^{-12} second. The violence of this change is really muted, however, compared to that which preceded it.

Next, when the temperature falls below $10^{15\circ}$, the vector bosons associated with the weak interaction disappear, at the epoch 10^{-11} second. Only photons remain as the fossil relic of these bosons. The photons, greatly cooled, are still around today: they are the famous cosmological microwave background radiation at 3°K.

As for the nucleons, formed from the remaining quarks, they were always accompanied by pairs of nucleons and antinucleons formed by energetic photons. However, as the temperature fell, the photons reached the point where they no longer had the energy associated with the mass of nucleons so they could not turn into them. At the epoch 1 microsecond the pairs annihilated, leaving only the residue of nucleons. The numbers of quarks of types u and d were equalised for reasons of statistical equilibrium, with the consequence that there were as many neutrons as protons.

The leptons, for their part, comprised electrons, positrons, neutrinos and antineutrinos, all in thermal equilibrium. Neutrinos particularly are extremely important in considering the ultimate fate of the universe even though they are among the least significant of particles: they have no electrical charge, their mass is zero (or next to zero, since the most recent measurements suggest less than 0.0001 the mass of the electron), and they interact with other matter scarcely at all and then only via the weak interaction. However, their abundance in the universe is such that it might give them an important role in determining its density and the expansion velocity.

At 0.5 second the density of the universe had fallen to the stage where the neutrinos became decoupled from other particles, on account of the feeble nature of the weak interaction. From that moment they followed the expansion of the universe independently, just like the photons of the microwave background, which did that at a much later time (300,000 years). At the time of decoupling they were in thermodynamic equilibrium with other particles; from that fact we can determine their density and convert it to the actual density today. The number density now is 100 per cubic centimetre. This is an enormous number; it means that neutrinos and antineutrinos rank as the second most populous constituent of the cosmos, just behind the microwave background photons which clock in at 400 per cubic centimetre.

At 0.5 second, the other leptons found themselves in a situation not unlike the nucleons and antinucleons at 1 microsecond: electron-positron pairs created from the radiation mixed with electrons remaining from the second detonation of the universe. These latter particles arose from the same asymmetry that X bosons caused for quarks. For this reason the residual electrons are present in exactly the same quantities as residual protons. That ensures that matter in the universe is electrically neutral.

Half a second later, photons could no longer replenish the electron-positron pairs being annihilated with new pairs. Only the residual electrons remained. Now the universe was one second old and had a temperature of 10 billion degrees. It was filled with photons, electrons, protons whose numbers equalled the excess electrons caused by asymmetry in the second detonation, neutrons in the same numbers as protons thanks to the quark u and quark d symmetry in that same detonation, and finally independent neutrinos and antineutrinos. The scene is now set for the beginning of the universe as we know it.

The major events during the first second of the universe give the impression that there was a lengthy process of churning, mixing, filtering out, and generally putting into good order all sorts of elementary particles created at the second detonation. At the end of this the only constituent particles remaining were those described in the 'First Look at the Universe'.

Because it set the conditions for all the subsequent fifteen billion years, we are quite justified in called this all-important first second the 'longest second'. Among the 10^{17} seconds since, not a single one of them had the same importance. And none will ever be that important in the future unless the universe contracts and collapses in on itself in a reverse of the Big Bang. In that case, the Big Crunch, the last second would be mighty important as a final goodbye. Actually, this possibility does not feature in the theory of the inflationary universe. The expansion is predicted to continue for eternity, with a minimal coasting velocity. The fate of the universe, thus written, goes very far into the future!

As for us, we each have 10^9 seconds at our disposal, and the one that will include our last heartbeat will have a significance like the longest second.

Time zero

Toward time zero

The intellectual challenges of the inflationary universe theory have opened unexpected vistas. We can be assured that they have led us to new ideas in cosmology, as well as about the universe in general (the inherent flatness of space, for example), and helped us build a plausible model for the first 10^{-32} second, up to the second detonation.

Have insuperable barriers been erected across our path? Hardly. All the same we must not underestimate the substantial difficulties. The main one is the rather rudimentary nature of the present theory: there are technical problems with the actual calculations, some brute force approximations, and arbitrary choices of parameters in the models. At this time there is no theory that we can follow with absolute confidence.

The main worry which theorists now have is that the cosmological constant, which determines the cosmic repulsion due to the vacuum energy, is observed to be tiny. It is extremely small in comparison to the value it had, according to the theory, when the vacuum dynamite inflated space. How the vacuum actually made its quantum leap from a state of very high energy to one with nearly zero energy is unknown. Some theorists consider that until that is explained, the inflationary universe is scarcely credible. Others think that physical processes as yet undiscovered will explain why the cosmological constant is now almost nil.

Logically speaking, we cannot say that with Grand Unification Theory we have achieved our goal. Our aim at the outset was not the mere unification of three interactions (strong, weak, electromagnetic), with the fourth (gravitational) being cast to one side. Supergravity, via supersymmetry, beckons us more and more. Regrettably, extensions of the theory are in a very embryonic state. Only a handful of physicists have got to grips with its complex formalism. The first monograph on this subject, aimed at researchers working at a very advanced level, was not even published until 1984. From these initiatives progress will now probably start to gather momentum.

The basic idea is that if we could track back to a time sufficiently close to zero, before the epoch 10^{-35} second when Grand Unification is valid, gravitation will be linked with the other forces of nature. At what epoch can we expect that to be true? Most physicists accept that it is at the Planck time. This time, which marks the unification of gravitation with relativistic quantum theories, is deduced from two lots of fundamental physical constants: one is the gravitational constant G, and the other Planck's constant h and the velocity of light c. With these three parameters there is a unique way of constructing a relationship with the dimensions of time alone, called the Planck time. The value is 10^{-43} second.

At the epoch 10^{-43} second, the universe was much hotter and denser, with a temperature of $10^{32\circ}$, than at the onset of the great inflation. The vacuum energy was 10^{20} times as large too. The first phase change of all occurred when the gravitational interaction became decoupled from the super-unified interaction. Right after that a standard expansion took place, driven by the pressure of particles and antiparticles, which were in thermodynamic equilibrium at a high temperature. As the expansion progressed, so this population of particles cooled, their pressure diminished, leaving at the first level the energy of the symmetric vacuum of Grand Unification. At the epoch 10^{-35} second we rejoin the inflationary scenario, leading to the second detonation, the longest second, and the great vistas of expansion.

Many physical processes must have taken place in the interval from 10^{-43} second to 10^{-35} second, even though at first sight that seems a negligible period. To see its significance, imagine a time interval of 10^{-45} second being represented by one year. In that case the interval being considered would be 10 billion years! Unfortunately the theory has not developed sufficiently for us to be able to say what took place during these earliest eons of the history of the universe.

Before time zero?

What took place before the Planck time? At the epoch 10^{-43} second the cosmological horizon extended to 10^{-33} cm, and the entire universe we now see was 10^{-55} cm across. This is unimaginably small, nothing compared to the cosmic apple just after the inflationary era. The entire universe existed in a sub-microscopic domain; it would have been under the control of the laws of quantum physics, subject to the uncertainty relations, just like an elementary particle. In particular, there can have been no sharply defined geometrical structure. Thus, if space was spherical our picture of its surface will not be a smoothly polished very tiny sphere. Rather we will imagine something puffy, with hills and hollows, tunnels, and maybe flecks breaking off; it would resemble the surface of a small blob of shaving foam, having the general appearance of a tiny ball but without a well defined surface.

Furthermore, this complex geometry was subject to incessant quantum fluctuations. Without any pauses it changed abruptly from one foamy surface to another in a disorderly manner. The detailed structure of space was not defined (just as in quantum theory there is no precise state for an electron). Only the statistical properties could be followed: at such-and-such an instant the puff of foam had an average diameter of so much, while in practice it fluctuated wildly about that mean. Or in some place on the surface of the ball the curvature had some average value; but again, from place to place, this curvature varied greatly, corresponding randomly to flat areas, needle-sharp points, wrinkles, and so on.

Physics under those circumstances gets reduced to the game of describing which states are more or less likely, just as with an electron we can only calculate the likely chance of its being here or there. It seems that the moment zero in the story of the universe will always be beyond our precise knowledge because of fundamental limitations imposed by physical laws. We will have to content ourselves with estimates and probabilities for this unique moment in time.

There are other obstacles preventing our reconstruction of the initial state of the universe, but they are of a different order. Suppose the state of the universe had been well defined at the Planck time and it had evolved subsequently in a way which could be precisely described by neat equations, and this had been satisfied until the inflation era. At that moment, catastrophe! The good order is totally destroyed when the universe heats to $10^{27\circ}$ when the vacuum energy is released. At such temperatures all physical structure is annihilated; only a high temperature gas in thermodynamic equilibrium exists, and this cannot furnish any information on its earlier states.

Information on the primordial state of the cosmos was destroyed by a fireball and this does limit our knowledge. On the other hand, this drastic incident does account rather well for the properties of the universe: isotropy, homogeneity, and so forth. The problem of the initial conditions, such as why did the universe start out isotropic etc, is solved What we have seen is that these conditions, assumed to be initial conditions, in fact emerged from the second explosion, independently of the real initial conditions at earlier times. This is a big plus for the inflationary universe scenario.

Nevertheless, physicists are not satisfied to leave it at that; is there not some possibility for a serious attack on the problem which everyone wants answering: 'what happened before time zero?'. This question has several slants to it: what was the universe like one hour before zero? Or, was there nothing one hour earlier? Or even better: did time exist an hour earlier?

According to relativistic models, the expansion could have emerged from a collapse, symmetrical with the Big Bang. If there were no quantum era between zero time and the Planck time, and none during the collapse, and if the equations of general relativity hold true during such a condensed phase, then in that case we could pose a further hypothesis to get a response to the question 'what happened before time zero?' The further hypothesis is that the collapse would be followed by a bounce and a new Big Bang. We could freely suppose, therefore, that our universe could have been preceded by a collapsing phase before the Big Bang. This scheme could also be invoked even if the present expansion will continue indefinitely; it would still have been possible for it to have been preceded by a symmetrical contraction extending infinitely into the past.

Such speculations are ruined by the quantum era, with its fundamental uncertainties. They don't let such a neat transition take place. Even worse, the actual theory doesn't permit any suitable transition. There are only preliminary calculations, made semi-classically, for models suggesting various possibilities for the evolution of space. According to these models, certain spaces starting with no dimensions some 10^{-43} second before zero time begin to oscillate a few times, thus getting their expansion underway, and then they merge with the classical solution. Other models, very large in the infinite past, contract, passing through very tiny but non-zero dimensions at zero time, and then branch into normal expansion. Yet other models, which are very large a little time before zero, suddenly contract rapidly and also go through a finite minimum at zero time, then finish as a normal expansion.

The list of possibilities is very varied. Even if theory progresses it will always have a certain air of mystery, because it is essentially probabilistic, when it confronts the problem of greatest metaphysical importance.

So far as the question about whether or not time itself existed before zero time, physicists and cosmologists know nothing of this. They have no elements for a response. That is not to say that such a metaphysical question should be left to one side. Absolutely not! The reason is that our earlier question indicated that time could have had no beginning and, in these circumstances, the question as to the existence or not of time before the Big Bang is very badly posed.

Metaphysical problems that scientists can ask sometimes have physical solutions, The semi-classical calculations prior to the Big Bang are one such example. Another, more striking, example goes back more than 30 years when, following the lead of George Gamow, physicists started to ask among themselves questions about the Big Bang. Many considered at the time that these were metaphysical issues, almost taboo subjects for physicists, and that it was unfitting to work on them. That state of mind only slowed developments in cosmology and that is a pity, con-sidering how the science of cosmology has subsequently advanced human knowledge.

Monopoles, strings and domain walls

A bar magnet is essentially a pair of magnetic poles, conventionally named north and south. A typical example is a compass needle. There are similarities between pairs of electrical charges, positive and negative, and the situation we encounter with magnets.

However, Maxwell's classical electromagnetic theory does not lead to complete symmetry between electrical and magnetic forces. This asymmetry means that isolated electric charges can exist in nature (such as the negative charge on one electron), but there are no isolated magnetic poles. If a bar magnet is snapped into two we end up with two bar magnets, each with its own north and south poles.

In 1931, at the beginning of quantum theory, Dirac combined Maxwell's electromagnetism with the new theory and proved that isolated poles could exist. From the electron charge e and Planck's constant h he even derived an equation for calculating the magnetic charge. Those two constants then intimately linked the electromagnetic and quantum theories respectively. The isolated poles, if they exist, are called magnetic monopoles.

The following stage had been reached in 1974, at the start of Grand Unification Theory. When the vacuum makes the phase change from symmetry to broken symmetry, defects in its spacetime structure may become apparent. They manifest themselves as a small cloud formed from all sorts of bosons: at the centre a core of massive X bosons 10^{-27} cm across, immersed in a much larger cloud of intermediate W and Z bosons 10^{-16} cm across, the whole lot swimming in a cloud of gluons and massless photons some 10^{-15} cm in size. Furthermore, this exotic mix is shrouded by a veil of fermions and antifermions out to 10^{-13} cm, comparable to the size of an atomic nucleus. The mass of such a topological knot in the fabric of spacetime is 10^{16} times the mass of a nucleon. These quantum defects have magnetic charge with the value Dirac predicted.

This then is a really curious particle: a layered structure, a total size about the same as some other particles, but with a colossal mass and carrying a single magnetic charge. It has hit the scientific headlines with the application of Grand Unification to the Big Bang because these monopoles ought to have been present in the cosmos in greater numbers than nucleons. Since that is not the case, the standard Big Bang model runs into a further difficulty.

Thanks to the inflationary universe, the situation is saved. Because of the enormous expansion at 10^{-32} second the monopoles have been enormously diluted to the point where they would be extremely rare in the observable universe.

Nevertheless the possible detection of one monopole has been reported. In 1982, Blas Cabrera of Stanford University announced detection of a single monopole using apparatus specifically designed for the purpose. The experiment searches for cosmic monopoles incident on the Earth's surface, using a small super-conducting ring. If a cosmic monopole passes through the ring it alters the electric current in the ring. In a period of six months just one such event was detected. Searches are going on because magnetic monopoles may bring us information about Grand Unification and inflation theory. They constitute one very rare example of our being able to test such theories.

Moreover, such exotic particles are of great interest both to physicists and to astronomers. They are massive enough to contribute to the density of the universe and thus influence its expansion. Possibly they could be trapped inside neutron stars by gravitational attraction, leading to their gradual elimination by accelerating neutron disintegration through catalytic effects. They could also be trapped in terrestrial rocks through bonding with ferromagnetic minerals. Some researchers have suggested that if this is so they should be released in the vicinity of blast furnaces, where millions of tonnes of ore are processed. Their own mass would make them fall, so they could be detected with instruments under the furnace sensitive to their magnetic charge.

Finally it is worth noting that monopoles are not the only consequence from the warping of spacetime during symmetry breaking. We have presented monopoles as pointlike defects, but there are other defects that are linear and they would appear like strings. These are the hypothetical cosmic strings.

Essentially, cosmic strings are linear regions of spacetime that differ from neighbouring regions because they have not been able to make the symmetry breaking phase transition. Their cross section is practically zero, but the energy of the symmetric vacuum that is locked into them is so enormous that one centimetre of cosmic string would have a mass of 10^{22} grams; that's the same as a lead sphere 100 km in diameter!

There could be two sorts of string: linear strings criss crossing the whole universe, and curved strings tied into loops. A loop of cosmic string a metre in size would have such a large mass that any object falling into its vicinity would be propelled by the gravitational attraction to several hundreds of kilometres per second, a respectable cosmic velocity. Such loops could trigger the density inhomogeneities that are essential for the formation of galaxies and clusters of galaxies. This would solve a problem that theorists working on galaxies have not yet resolved so far.

Linear strings, by contrast, do not have gravitational attraction. Nevertheless they would deform spacetime in a special way. To picture this we shall imagine the two-dimensional space defined by a plane perpendicular to the string. Then a model of the curvature due to the presence of a string is furnished by the surface of a cone (not the sphere nor the saddle back topology we encountered earlier). The geometrical situation can be described by imagining cutting a sector from the plane, like a slice of pie, and then folding this so the two edges join (figure 18). Now a complete circuit round the string takes less than 360°, which would be the value for a round trip in Euclidean space. If you took a gyrocompass on such a trip it would get more out of register on every circuit!

Euclidean geometry is valid for the surface of the cone because we made it simply by gluing a flat sheet of paper. The space is not, strictly, curved and no gravitation is felt anywhere on it. Only the very peak of the cone is curved, and strongly, so there is strong gravitation on the string.

This sort of space, which we can call conical space, affects in different ways rays of light travelling on either side of the string. For that reason, if such a string crossed the heavens it would slightly alter the strength of the microwave background radiation from one part of the sky to another. It would be possible, in principle, for it to double the image of a quasar situated beyond it.

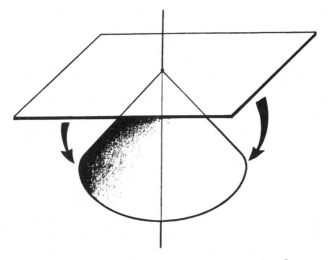

Figure 18. A cosmic string changes the geometry of a plane perpendicular to its length into a cone.

The twinning, or gravitational lensing, of quasar images has already been observed, but is thought to be due to intervening galaxies rather than cosmic strings in the actual cases to date. All the same, very precise measurements need to be made of the 3° radiation in the neighbourhood of these double quasars just in case we can confirm that they really are being caused by cosmic strings. You can see that linear cosmic strings have truly extraordinary properties and the study of them is still in its infancy.

Cosmic strings are held under enormous tension, because the false vacuum exerts negative pressure. This fantastic tension makes them vibrate, zoom around at the speed of light. However, because the diameter is miniscule, they have practically no direct effect on anything that they encounter.

In contrast, the conical space generated by the linear strings leaves behind a wake of spacetime anomalies. As a thought experiment we can consider what would happen if we encountered these cosmic strings. If one zipped through the room you are in, you would certainly notice it: the walls on either side of the trajectory would start to move towards each other with a relative speed of 4 km/second. However, unless the string also passed through your body you would not feel much, apart from a sudden fright until an object to one side were suddenly thrust straight at your face! If, however, you were walking through a string, you would see each side of your body approach the other.

These astonishing fossil relics from the Big Bang are mixed up with cosmic domain walls. To picture these we need to go back to the transition from the false vacuum (symmetric) to the real vacuum (broken symmetry). Let's think about water freezing into ice again. In general, the water does not solidify into a single crystal, but into a matrix of little crystals jumbled together; each separate crystal has a different orientation to its nearest neighbours. We can consider the crystals to be separated by walls that mark the transition between their differing orientations. When we consider the vacuum phase transition, the equivalents of these boundaries are defects in two dimensions, termed 'domain walls'. They separate regions of the universe that have different properties on either side of the wall. At the present stage the differences are not manifest. But, let us make a flight of fancy: then the differences will surpass those we would experience if we could suddenly transfer ourselves to another planet, just like in the wildest genre of science fiction. In all probability it would be catastrophic for the observable universe, whose scale is much smaller than that of a cosmic domain, to cross a domain wall. Theorists speculate that the traverse would have to be made at the speed of light. Then, the catastrophe would be instantaneous and, even worse, there would be no advance warning of any kind. On one occasion at a conference, Guth presented a table giving the main stages in the evolution of the universe from the Planck time to the moment of his presentation. He stopped the table at that point, not wishing to prejudge the future, making the comment that the vacuum could be metastable, in which case time as we know it could be snuffed out at any moment. This sudden and radical destiny for our universe is something to muse on, perhaps...

Needless to say, these weird 'heavenly bodies', magnetic monopoles, strings and domain walls have excited a lively debate among physicists and astronomers. Many major international conferences have been devoted to discussions of the ideas outlined here. It is impossible to say where this study will lead to in a few years' time. It certainly is possible that they could be relegated as also-ran ideas if observations are in conflict with them

The universe of universes

Countless universes

At the end of inflation, at time 10^{-32} second, today's observable universe was as big as an apple, although the vacuum 'crystals' at symmetry breaking were 10^{25} times as large, reaching 10^{26} cm. The extent of the crystals at that early time was 1000 times as big as the scale of the observable universe today. Thus our universe is immersed in a domain of staggering immensity.

This comparison means that the cosmological quest is not even simply a matter of knowing the totality of one domain, ours, alone. We should not ask then what can be learned about other domains. The inflationary universe scenario teaches that our universe is but a miniscule part of a much larger and homogeneous whole (our crystal), and that in its turn that is merely one large entity in a cosmos of many universes. In that sense there could exist innumerable universes, according to the inflationary picture.

Unhappily, there is nothing in the theory, and even less in observational terms, that will give us a window on these parallel universes, not even to the extent of confirming that they exist. Is the road to further knowledge on this front completely blocked? Perhaps not. First of all, with the passage of time the cosmological horizon recedes, and in 10 billion years time it will be much farther away than it is today.

Secondly, progress on the theoretical side should give a more precise model of the part we can see. In mathematics there are functions that can be extended beyond the domain in which their properties are known; so, perhaps it is not unthinkable that we could extend physical theory to other universes. For example, it is already the case that certain properties of the 3°K radiation, considered on the largest scales, have implications beyond the horizon.

However, future progress will probably require the incorporation of quantum gravity, and as we have seen, in the quantum world we are only allowed to calculate probabilities. So the theory will run into the same fog as the theories dealing with the time before the Big Bang. We can give an arbitrary and speculative example of answers to the questions we might ask as to whether other universes could resemble our own: in the countless universes of universes, a fair fraction will contract, certain ones will have no matter, others no stars, very few will have carbon, and even rarer still will be universes with organic macromolecules. And, exceptionally few universe will contain life based on water and carbon chemistry. And, on top of this, evaluations of probabilities may eventually be produced.

Even though these results are merely statistical, they are of fantastic interest, particularly because of their philosophical implications.

The ultimate free lunch?

Perhaps the first and most striking idea to emerge from the history of the universe is its sudden birth: the cosmos came from the Big Bang 15 billion years ago. Solid theory, underpinned by high quality observations, supports this position. General relativity accounts for the explosive origin; quantum physics explains the unannounced appearance of particles. Beyond these theories, still speculative notions endow the vacuum with amazing properties from which the inflationary picture is assembled: the cosmos emerged from the vacuum itself! These recent developments in cosmology constitute the second striking notion in the story of the universe.

It is only a small step to float from these ideas to the robust statement that the universe appeared out of nowhere and out of nothing. As a matter of fact, so far as physicists and

astronomers are concerned the idea that the universe started
as a quantum fluctuation out of absolutely nothing dates from
1973, when Edward P. Tryon of Hunter College, City
University of New York, formulated the idea from the
following facts.

First, the net electrical charge in the universe is zero,
with every proton being balanced by an electron, the other
particles being neutral.

Second, the universe is not rotating to any significant
extent. Of course, almost all celestial bodies are turning on
some axis of rotation: planets, the Sun, stars, galaxies, and
clusters of galaxies all have some spin or rotation. But the
rotation axes on the largest scale have random orientations so
that the net rotation in the universe is nil.

Third, and this is the novelty of Tryon's ideas, the total
energy content of the universe could be zero too, despite
appearances to the contrary. In the cosmos we have both
matter and radiation, both of which represent forms of energy.
Then there is also the kinetic energy of every kind of celestial
object, the recession of galaxies, and the expansion of the
universe. We can lump all these together as non-gravitational
energy. Now consider the energy locked into gravitation. Were
the universe to contract and fall in on all the matter that it
contains, a considerable amount of energy would necessarily be
released as objects tumbled literally onto each other. Clearly
they would release gravitational energy in such a process.

When each side of the equation is computed. Tryon
found that it turns out that the gravitational and non-
gravitational energies more or less cancel each other out.
Therefore the total energy content could be zero. And this
basically is his hypothesis: the universe has no electrical
charge, no rotation, and no energy, so it could happen out of
nothingness, this nothingness also lacking charge, rotation and
energy.

Grand Unification and the inflationary universe can go
one better than this: the nucleon number in the universe is not
necessarily constant because of the possibility of proton decay.
Hence, the universe could have an initial state with no

nucleons, and no matter. Such a universe, without matter, could also emerge from nothing — also without matter. However, physicists consider that the symmetric vacuum is not necessarily the 'nothingness' of Tryon's hypothesis. Even though it is a vacuum, it has properties, which are still poorly known right now.

Such ruminations have their philosophical side. To paraphrase remarks already made by Guth: there is nothing about the universe that markedly distinguishes it from the vacuum; this being so, it is tempting to think that the universe began from nothing. The inflationary universe certainly illustrates this as a possibility. A common cliché is that 'there is no such thing as a free lunch'; nevertheless it now seems possible that the universe itself is the ultimate free lunch.

Maybe this perspective suggests that in this speculative part of our journey I have brought you on a wild goose chase. All of this chapter is very tentative indeed. The theory will surely grow old fast and there will be important adjustments to be made. All the same, it gives me great pleasure to attempt to communicate the enthusiasm and intoxication of this wonderful research. These last years have been exciting ones in the search for the primordial secrets of the universe. In my approach I have left technical problems largely to one side, being less interested in the research itself than in the large scale perspective it gives on the ultimate fate of the universe. For example I did not dwell on things like the formation of galaxies, and the problem of missing mass in galaxies and clusters of galaxies, which are in rather preliminary form anyway, but from which new insights might spring forth.

7

The universe and ourselves

The trail to life

In the last chapter we said that the Big Bang and the appearance of the universe from the vacuum were two very striking facts. Now we want to turn to a third, very striking, impression, which emerges from this long history of the universe: the fate of the universe seems quite fantastic. The destiny of the universe seems mischievously entwined with a lot of disconnected events. Some of these took place unimaginably quickly, such as the burst of activity during the first 10^{-32} second of inflation, or the sudden intervention of a cosmic domain creating havoc in its path. Others seem rather protracted affairs, such as the lethargic progress that followed the first quarter of an hour, and the launching of the universe on a never-ending expansion.

Key events like these arise from microphysical properties, such as the Heisenberg uncertainty relations, and from processes operating on the largest scale, such as the expansion of space as described by Einstein's equations.

Among the decisive events, some had direct action on the course of the universe and, therefore, on our place within it. The eventual disappearance of matter, through proton decay, would have major consequences for the universe and in particular for us: no matter, no humans.

Then there are yet other events, which seem rather trivial so far as the cosmos is concerned, but for us they are crucial. Take the nucleus of a carbon atom for example. This has energy levels and one of them is very precisely defined in the following sense: if it were only a little different, the amount of carbon made by nuclear reactions inside the stars would be very much smaller than it is; the universe could cope with this without being very much different, but without carbon there could be no organic chemistry, no life, and therefore no humans.

Since these primordial events are of such seminal importance to humans, we should pause and consider them. After all, we are an integral part of the cosmos. Above all, we are a component of the universe that will not necessarily always remain passive. Nobody can assert that none of the conscious, thinking, beings in the cosmos that possess technology will not affect in some way the destiny of the universe. The destiny of our own planet, for example: we, for our part, have already modified that to some extent have we not?

Planetary systems

We live on a planet orbiting a star, so the notion of planetary formation in the universe is important as far as we are concerned. What do we know about planetary systems? Better still, are there any planets like our own? These are basic and straightforward questions without clear answers.

For decades astronomers have searched for planets by photography with powerful telescopes but the results are unconvincing. To detect a planet orbiting even a nearby star is very difficult. There have been hopes dashed over the years. Serious progress will have to await the advent of the Hubble Space Telescope and new detection techniques now under development at several observatories.

Theoretically we can state that, when a star forms by condensing from interstellar gas, some material will orbit the protostar as a disc. Subsequently this proto-planetary disc can fragment into blobs, which then join up to form planets. A huge amount of observational data accumulated in the last 20 years or so is entirely consistent with this idea, without actually proving it.

A simple argument would seem to point to the existence of other planets. Since the Sun has planets, and since it is a common run-of-the-mill star at that, other stars of similar type ought to have planets. However, it is possible that the Sun acquired planets through some extremely rare process which has only the slightest chance of occurring a second time.

In the last few years, new infrared telescopes have produced encouraging results on the existence of other planetary systems. For example, it seems very likely that there is a disc of material encircling Vega, although more powerful observations are needed to confirm this. About half a dozen similar cases have been reported. It is possible that they are proto-planetary discs which, in time, will condense to form planets.

In 1982 the International Astronomical Union created a Commission for the study of Extraterrestrial Life. One of the main priorities it has listed for the next ten years is to search for planetary systems in the vicinity of the nearest one hundred stars. Then perhaps we will know whether planetary systems are common in the universe. That will have important repercussions for the big question: are we alone in the universe?

Inhabitable planets

If life more or less like that on Earth is to develop in a planetary system, then there need to be reasonably hospitable planets within that system. The whole issue of planetary evolution and the possibilities for life to emerge is another of the priority items on the agenda of the IAU Commission for the study of Extraterrestrial Life.

We have already made spectacular progress by means of direct space exploration in our own solar system. Completely new frontiers have been opened: there are nine planets with 44 natural satellites, and 20 of those are worlds more than 200 kilometres in diameter. We can add over 3000 minor planets, or asteroids, and 30 of those are over 200 kilometres across. Altogether there is a respectable armada of 60 significant objects cruising through the solar system.

Which among them could sustain life? If economic activity continues at an appropriate level, then it is quite probable that in less then a century from now humans, even in communities, could be living on a fair number of those worlds, using the local resources for energy and materials. So, these worlds do have the capacity to support life, admittedly at a rather advanced stage of technology.

Taking the concept of 'natural' life (a subjective and artificial notion anyway), few of the 60 globes are suitable for it. Earth, the planet of oceans, the blue planet, is the most suitable. It has an abundance of water, something considered to be of major importance for the appearance of life. Liquid water sets tight constraints on physical conditions because the ambient temperature must be between 0° and 100°C at atmospheric pressure. There is a high probability that this has been so for billions of years, without a break, on Earth. No other world in our solar system has been so blessed, as evidenced by Venus and Mars, the only other plausible candidates.

Earth also has massive amounts of oxygen in its atmosphere. Oxygen is, in fact, a by-product of living systems. In the course of billions of years, organisms using photosynthesis have released it from carbon dioxide. The blueness of our planet is not, therefore, intrinsic, but is a consequence of life. We must hope that mankind will not destroy this oasis in space; for as long as possible it should be a pleasant abode for life.

Let us now consider the Earth's nearest neighbours, Venus and Mars. Venus, which is somewhat closer to the Sun, is rather warmer, and because of that it hasn't managed to rid itself of the primordial atmosphere of carbon dioxide like the Earth has. This carbon dioxide acts like the glass in a greenhouse, trapping the heat from the Sun. Venus is like a hothouse with all the windows firmly shut. The result is that it has no liquid water, and a temperature at the surface of 450°C. In all probability it has never had life.

So far as Mars is concerned, the prospects seem a little brighter. It is further from the Sun than Earth. Geological studies show that there was abundant surface water, three billion years ago. In particular, Mars has ancient dried river beds, whose output must once have been like a thousand Amazons! Since that happy time, Mars has been in the grip of an ice age. There are irregularities in its orbit round the Sun, due to interactions with other planets. These, together with the photochemical effects of solar radiation on its atmosphere, have caused it to lose almost all of its protective blanket and freeze to sub-zero temperatures. The remaining water is frozen as a permafrost beneath the surface. The two Viking landers of 1976 found no indication of life in the top few centimetres of soil analysed at the landing sites. But this does not discount the possibility that life could have appeared on Mars in its first billion years, as happened with Earth, and could now be dormant in the depths of the permafrost. To examine a question as fundamental and as important as this we must return to Mars, and several long term missions, which could culminate in manned landings in the early part of the next century, are now underway. That two planets that are almost cousins could have such different histories shows how the development of life on a planet must hang in the balance.

Some meteorites are fragments of asteroids, and these furnish further material for research on the origin of prebiotic molecules. Uncontaminated meteorites have been collected in recent years from the ice fields of Antarctica. The same precise methods of analysis as used for the lunar samples of the Apollo missions have detected amino acids of non-terrestrial origin in some meteorites. Amino acids are complex organic molecules that are used to build protein molecules. Their presence is an important leading indicator of prebiotic activity, at the time when the planets were forming, in the first parts of the proto-planetary nebula to condense: dust grains, meteorites, and asteroids. Possibly comets too have preserved evidence of ancient prebiotic activity, something that only space exploration could confirm.

Saturn's largest moon, Titan, is also a promising candidate as a seat of prebiology. Its diameter is 5000 kilometres, the surface pressure one and a half times that on Earth, and the principal gas in the atmosphere is nitrogen; these characteristics make it more like Earth than any other solar system body. The action of solar ultraviolet radiation on its atmosphere synthesises complex organic molecules, such as methyl acetylene. The most interesting of them is hydrocyanic acid (hydrogen cyanide), because five of these can be joined under laboratory conditions to make adenine, a fundamental component of the double helix DNA, which is the basis of heredity in living systems on Earth. In the double spiral of DNA there are pairs of nitrogen-bearing bases, and adenine is one of the four types.

Significant prebiological activity could be present on Titan, but because the temperature is low (180°C below zero), this cannot have led to true biological lifeforms. Titan is a model of the primitive Earth, preserved in a freezer. Soon after AD 2000, it is possible that Titan will be explored remotely. Information on its prebiotic state will tell us something of what stage the Earth went through on its route to the development of life.

Exploration of the solar system raises problems and questions about how the origin of life, and therefore of ourselves, fits into the fate of the universe. Will spacecraft journeying to Mars, Titan, comets and asteroids before the next millennium give us some answers?

Life on Earth

The first solid objects were formed from the proto-planetary nebula about 4555 million years ago. Hundreds of millions of years later the Earth condensed; initially it was liquid, because of the heating effect due to the liberation of kinetic energy by cascades of planetoids and meteorites raining down on it. These bombardments became progressively less intense, the temperature fell, and 3800 million years ago the continental plates solidified; next, the oceans condensed from water vapour held until then in the atmosphere.

How did life begin? That is still an enigma. The facts are that the oldest rocks, from 3500 million years ago, contain fossil organic molecules; in particular, the ratio of the isotopes carbon 12 and 13 is different to that in minerals. The suggestion is that a flourishing biology developed rather quickly. Where from? That is the big unknown. Laboratory simulations of the primitive physical and chemical conditions at that epoch are readily able to produce amino acids, nitrogen-bearing bases, and sugars which constitute the building blocks of life. But they are only bricks, and they lack the machinery for assembling life itself.

A billion years later, photosynthesis was 'invented' and oxygen production commenced. At this time living organisms were single-celled bacteria or prokaryotic cells which existed in three principal varieties. These small cells were about a micron (a ten-thousandth of a centimetre) in size.

Around 1400 million years ago a major breakthrough took place: the appearance of eukaryotes, cells that contain true nuclei. Intermediate stages of evolution must have occurred in the transition from prokaryotes to eukaryotes, because the latter are one thousand times larger; they are complex 'factories' with several specialised 'machines': there is a nucleus in which DNA-containing chromosomes are enclosed by a membrane, organelles called mitochondria for respiration, chloroplasts for photosynthesis, Golgi bodies for excretion, ribosomes to synthesise protein, ...

Eukaryotes ruled the planet for nearly a billion years, flourishing in the oceans and ceaselessly working on the release of oxygen. Around 670 million years ago a giant leap forward occurred. The new ecological niche, oxygen rich, favoured the appearance of the first multicellular life, and it dominated for the next 120 million years or so. This was the Ediacaran period, just at the beginning of the Cambrian era. The name comes from the Ediacaran Hills in the Flinders Ranges, South Australia, where this life demonstrably below the Cambrian was recognised.

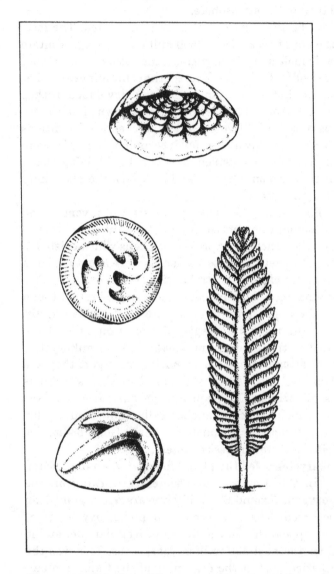

Figure 19. Primitive life, preserved as fossils from the Ediacaran period 600 million years ago.

The way in which they are preserved as fossils suggests that they were stranded on muddy shores by retreating tides and then buried in sandstone deposits. The marine creatures of that period were soft-bodied and flattish, thus presenting a large surface area to the water, which contained the oxygen they needed for metabolism (Figure 19). There were roughly thirty species (four phyla) looking somewhat like jellyfish, primitive slugs, and soft corals with a feathery appearance.

Then, around 550 million years before the present, came the Cambrian explosion, with its remarkable diversification, particularly of animals. Even the wildest science fiction can scarcely hold a candle to this fertile era, which was real, not fiction. Suddenly life itself wanted to participate more fully in the world around it. We can get an insight into the extravagance of Cambrian design just by listing some of the thirty phyla of animals known from that time, each one of which requires a different structure or metabolism in order to live: sponges, sea anemones, worms, insects, starfish, squid and octopus,..., and the vertebrates.

Vertebrates exhibit great variety among their different classes: bony fish, reptiles, birds, and mammals. Mammals in turn can be split into ruminants, carnivores, marsupials,..., and primates. Among the primates we think especially of the apes, because they are our next of kin in evolutionary terms. And they for their part are very varied: mankind, gibbons, orangutans, gorillas, and chimpanzees.

At the highest level in this evolutionary tree we find mankind and intelligence. Mankind is well-endowed with intelligence, but does not have a monopoly of it. Genetically, chimpanzees are more closely related to mankind than to orang-utans; but for a few tricks of evolution it could be chimps sitting at personal computers with a big grin on their faces, while we languished in zoos crying out for our intelligence to be appreciated.

In large jumps the above is a brief outline of how life has developed on Earth, from simple origins to the highly complex concepts of intelligence and human consciousness we see today. On this insignificant planet in the immensity of the universe, this intelligent life has only emerged in the last few million years, which is just a twinkling of the eye on the ten billion year timescale of the cosmos.

This line of thought can also thrust us hundreds of billions of years into the future. It is scarcely credible that our intelligence is the pinnacle of the potential of the universe. To think that would be unfruitful, damaging, and above all, unscientific. There is no rational indication that would allow us to assert that human intelligence is the most advanced form of intelligence in our particular bailiwick.

The future seems too extensive for that, although catastrophes and mass extinctions seem to overwhelm everything in the long run. Life on Earth, for example, has been subject to extinctions; there was one 65 million years ago which wiped out a good half of marine species; and one half is not negligible. At that time the demise of dinosaurs opened the way for the rise of mammals, and thus of hominids and mankind. There will be irreversible catastrophes in the future: the expansion of the Sun when it becomes a red giant star, supernova explosions in our vicinity, collisions of the Earth with asteroids..., temperatures near absolute zero in indefinite expansion, or the inferno of the final contraction.

Unfortunately we cannot possibly extrapolate from our present condition to some future state of higher intelligence. Wild speculations based on nothing and with no hope of verification can scarcely help us to make progress.

However, one of the priorities of the IAU Commission on Extraterrestrial Life could, in principle, give us information on higher forms of life in the universe: this is the search for and eventual detection of artificial radio signals; if successful this venture would give information on more advanced forms of intelligent development in the universe. Quite possibly the Earth is a perfectly ordinary planet among innumerable other planets; this uncountable number could be as big as 10^{22} in the observable universe. Furthermore, the Earth has formed sufficiently late in the history of the Galaxy for there to be planetary systems formed billions of years before us and, therefore, having had the chance to push the development of life much further.

The anthropic principle

The history of the universe seems like a succession of lucky breaks in some gigantic lottery which have been packaged in such a way as to give rise to the particular type of universe in which life is possible, rather than choosing other routes which would result in any kind of universe we care to imagine.

The amazing way in which the universe has evolved makes the discovery of any kind of cosmic order or purpose rather difficult. After our very careful consideration of all the chancy rolls of the dice which preceded mankind, it is tempting to say these two things: firstly, a particular importance can be attached to our emergence and, secondly, to say that the seemingly random dice throws were in fact directed straight at our appearance.

This fatalistic view reminds me of Bernardin de Saint-Pierre's whimsical remark: 'pumpkins are big so that entire families can eat them'. Fred Hoyle has commented in the same vein: 'a commonsense interpretation of these facts suggests that a high intelligence has cheated the laws of physics by somehow grabbing the winning cards'. Here we have a vision which is at once absolutely free but also sterile, because it cannot be verified. It is also a very extreme view, for two reasons. Firstly, it attributes a degree of importance to mankind's origins which definitely doesn't emerge in the results I have presented so far in this book. And secondly, there is a less extreme and more realistic view, which has been expressed thus by Brandon Carter: 'the universe should be so constructed as to admit the creation of observers from within itself at some stage'. This statement takes account of the fact that observers, humans, are here. This formulation of the universe and mankind's place within it is known as the anthropic principle.

Philosophically the strong anthropic principle can be reduced to the statement that the universe is here to be lived in, which strikes a resonance with some religious views: God made the world for mankind to live in. John A. Wheeler and others have even gone as far as to say that our existence is the cause of the special structure of the universe. This last point is rather obscure because a causal connection from humans towards the universe does not seem to have been possible in the past.

The fundamental starting point is this: 'the existence of any sort of organism whatsoever that can be described as an *observer* is only possible if the universe has certain constraints' (B. Carter). Or, (J. A. Wheeler): 'mankind is here, so what should the universe be like?' From this we can derive a weak statement of the anthropic principle (B. Carter): 'those things which we ourselves are now able to observe are constrained by the fact that we are here as observers'.

The weak anthropic principle does not go as far as the strong statement, in philosophical terms; mankind is no longer the purpose of the universe, but is an observer of it. This statement conforms to the facts and expresses the anthropic 'principle' as a factual statement instead.

This is a healthy situation, for it leads to a question which we can attempt to answer. The beginnings of a solution to the conundrum were formulated in 1957 by H. Everett, who made the suggestion that we should accept all possible alternative universes as being real. Under those circumstances the solution to the weak anthropic principle would be similar to the fact that the surface of a nice planet had been selected for us out of a vast collection of less habitable cosmic places (Paul Davies). Who is out there to be surprised that we live on the Earth and not Venus? By applying the weak anthropic principle to the manageable scale of the nine planets of the solar system, we see how its solution works.

A priori, any attempt to transpose this local example to a vast array of different universes poses problems of a different order. But now, like a magic charm, we can once again roll out the inflationary universe. It sometimes does happen in science that everything falls magically into place, just like the last few pieces of a jigsaw slot together effortlessly. But we must still watch out! Until everything is in place we haven't finished the game.

The various 'crystals' that resulted when symmetry breaking took place, and when the universe made the transition from the symmetric vacuum to the vacuum with broken symmetry, are like many different universes. A. D. Linde has commented that many universes could exist, each disconnected from the others, and we happen to live in just one of these. The simplest way of taking account of the anthropic principle is perhaps via the inflationary universe scenario. All the other 'desert islands' on which other forms of life could in principle flourish are then of no importance so far as we are concerned. The inflationary universe scheme can thus serve as the foundation of a type of weak anthropic principle.

The last words on this, also from Linde, are perhaps the best: 'According to Guth the inflationary universe is the unique example of a free lunch, because in this concept all matter is created from the unstable vacuum. Now we can add that it is the only free lunch where all the items on all the menus are available.'

The far future

Current views suggest that the nature of the universe in the far future will depend crucially on whether the expansion of the universe will continue indefinitely. Will the universe drift down into a cold vacuum or will it return to being hot and dense? In the next few years this question will be answered as a result of careful analysis of data from the Hubble Space Telescope in Earth orbit. By the time it is up and running, this telescope will have taken twenty years from conception to operation, partly because the launch was greatly delayed by the Challenger accident. In that time a whole generation of astronomers has worked towards its success. Cracking the ultimate secrets of the universe is not easy. Even with the Space Telescope it will need a great deal of observing time and international collaboration.

I have personally devoted many years of my scientific career to a specific aspect of the problem: measuring the distances to galaxies. This is important data since the ratio of velocity to distance gives the universal rate of expansion, the famous Hubble constant, H. To find the distance to something as remote as a galaxy is hard work. Suitable surveyor's chains don't really exist for such large distances. Instead, indirect methods have to be used, by seeking distance indicators in galaxies whose distance is known independently and then applying the technique to more remote objects.

To do this I used novel techniques in radio astronomy. The then new radio telescope at Nançay, France is one of the two or three most powerful in the world. It was used to examine the 21-centimetre radio waves emitted by interstellar hydrogen atoms. These radio signals were carefully recorded, reduced by dint of hard work far into the night, accumulated, sorted and correlated, analysed, and mercilessly sifted by all sorts of statistical techniques. Then they were linked to observations made in visible light so that bit by bit we can establish mileposts through the hitherto uncharted reaches of deep space. Listening carefully to the radio waves, the universe seemed to whisper this result: H = 25 kilometres per second for every million light years of distance.

At the present time other groups using different approaches get other values: H = 15, or 35,... Who is right? It is still a mystery. We have all run into a brick wall: to leap over it we are going to have to do something pretty dramatic, with fresher and more powerful tools. This is what everyone hopes to get out of the Space Telescope. Because the far future of the universe depends on the value of the Hubble constant.

The contraction option leads to a relatively limited future. The ultimate inferno could engulf us in a hundred billion years. By contrast the option of indefinite expansion opens the way to an infinite future. Recently theoretical physicists and astronomers have asked: if time does not have some limit; what fundamental limits will the cosmos itself eventually confront?

Let us look at how things will appear in the future. Clusters of galaxies will then be getting further and further away from us, because of the expansion of the universe, while they continue to hold together the individual galaxies of their own gravitational empire. Despite this, more and more clusters will be observable because the cosmological horizon is shooting away faster than the expansion of space. What this means is that we shall become more and more isolated (expansion) and less and less alone (more universe can be seen).

At some point we shall reach the first day of reckoning when every star in every galaxy will have ceased nuclear reactions and thus the production of light and heat. A good fraction of their atomic nuclei will be in the lowest possible energy configuration, existing as nuclei of the element iron. Stars deprived of the means of sustaining thermal equilibrium must contract under the inwards pull of their own gravity. The least massive stars essentially become spheres of iron sustained by the pressure of an electron gas. They just get cooler and cooler, becoming 'black dwarfs' about as large as the Earth. Somewhat more massive are the neutron stars with diameters of about 10 kilometres. In these stars the electrons and protons have formed neutrons under the influence of the gravitational pressure. Finally, the most massive stars overwhelm even neutrons and they cannot halt inexorable uncontrollable gravitational implosion. These stars become stellar black holes with diameters of a few kilometres.

This first stage takes about 10^{14} years, which is a timescale one thousand times greater than the hundred billion year future in the hypothesis of a contracting universe. Over that time galaxies will become faint systems composed entirely of black dwarfs, neutron stars and stellar black holes. Planets, asteroids and dust will still exist because the interatomic forces are able to resist the self-gravitation of such small bodies.

Let's move on to stage two: planets leave their parent stars because they are dragged by tidal forces of neighbouring stars which, in the course of time, just happen to breeze by. The timescale for this epoch is 10^{17} years.

Next we get to stage three; what happens there? Ninety per cent of stars get wrenched from their parent galaxies by tidal forces (with other stars) and they are left to drift in intergalactic space. The ten per cent remaining, by contrast, get concentrated progressively towards the galactic centre. Frequent collisions make them coalesce into a highly condensed mass, leading to the formation of a galactic black hole which swallows all of them. Quite possibly this is similar to events which we think have already really taken place in quasars.

Within the nuclei of quasars and active galaxies there is a central engine, producing energy on a stupendous scale. This source of activity is responsible for beaming electrons and magnetic fields up to a million light years away from the nucleus. The most plausible model for this activity is one in which matter is falling into a very massive black hole. This process increases the mass of the black hole, of course. Careful studies of the orbits of stars very close to galactic nuclei suggest that black holes of up to one billion solar masses are present in some objects. These supermassive black holes have dimensions of order one light day. The epoch of this third phase can only be guessed at: maybe 10^{18} to 10^{24} years.

Towards the end of this distant era, galactic black holes in clusters of what used to be galaxies start to interact gravitationally through close encounters within the cluster. For the most part they disperse to intergalactic space because they get flung out by a type of gravitational sling shot effect. Those that remain fuse into a 'supergalactic black hole'. They contain hundreds of billions of solar masses and are about one light month in size.

So, by around 10^{24} years from now the universe is sprinkled with supergalactic, galactic and stellar black holes, black dwarfs, neutron stars, planets, and dust, all in flight from one another through the gloomy darkness of a vast universe. This disheartening situation continues for a very long time until new, but very rare, phenomena start to come into play.

Black holes

Consider a ball of matter like our Sun curving space in its vicinity. We can visualise this curvature as a sheet of stretchy rubber with a heavy ball resting on. Where the ball rests a depression is formed. This gets deeper if a ball of given size is considered to be made of a denser material. There exists some (large) density at which the curvature of the sheet snuggles up to match exactly the ball's curvature. Space has then turned on itself. It is limited to a spherical horizon shrunk wrapped round the ball. If the density of this ball is now increased further by compressing it the horizon remains in the same place, for its dimensions depend only on the mass of the sphere. Such a rubber sheet horizon has properties like the cosmological horizon.

A point mass falling towards the horizon will seem to have a larger and larger redshift the nearer it gets, but it never actually reaches the horizon! A movie of its fall will seem to run slower and slower as the point nears the horizon. In marked contrast to this consider an observer falling to the horizon: the time needed to reach it is very brief, as perceived by the falling observer.

These bizarre effects are a direct result of the complex curvature of space as described by general relativity. The physical conditions inside the horizon are not at all well known. All we can say for sure is that the mass is confined within it because no message can pass from the interior to the exterior; since its behaviour external to the horizon depends only on the mass, the exact nature of the matter within it does not concern us.

The cosmological horizon is the inner surface of a huge sphere which surrounds us on all sides. The horizon of a black hole is a tiny sphere shrunk around a point in space, and we can only observe this spherical region form the outside. Since we cannot penetrate it, the meaningful dimension to think about is the circumference, which we could at least travel around, in principle. For Earth this circumference is about 5 cm, and 18 km for the Sun.

The first of these would be the disintegration of nucleons, which is implied by Grand Unification Theory. This starts to happen at around 10^{32} years, some 10^8 times later than the previous era. The majority of the nucleons are inside black dwarfs, stars, and planets. Their decay releases energy which might slightly warm up the icy stars to maybe a couple of hundred degrees above absolute zero. The rest of the nucleons are in the intergalactic medium where their decay will provide sparsely distributed electrons and positrons. What is meant by sparse in this context? Their average separation will be something like the diameter of our Galaxy today, 100,000 light years. Despite these immense distances they will form, in time, huge atoms of positronium in which an electron and a positron are in mutual orbit. Given even more time, the orbits decay, the two particles get closer and closer, and eventually they touch and are thus annihilated in a puff of electromagnetic radiation.

This fourth distinct scene in our story of the far future is marked by nucleon decay. It is a scene which will be fatal to all forms of life and it sounds like the death knell of the cosmos. Furthermore, this endpoint could arrive quite a bit earlier if nucleon disintegration is catalysed by magnetic monopoles, the *enfants terribles* of the inflationary universe. Everything will depend on their space density. We do not know what this might be but various theories give upper limits. Typically a neutron star could be eaten away by the monopoles trapped inside it in only 10^{11} years. The Earth itself would flake away in about 10^{18} years. Will life need to extract the monopoles in order to delay this process?

While musing in this fashion we must remember that proton decay has never been observed. If nucleons are, after all, stable then matter will endure. But for how long? On the longest timescales it is still threatened. Now we can introduce a variation on stage four: the disintegration of matter via virtual black holes. This surprise package is courtesy of one of the greatest exponents of the quantum effects of black holes, the Cambridge physicist Stephen Hawking. According to him, quantum fluctuations make it possible for quantum black holes the size of the Planck length (10^{-33} centimetre) to make brief appearances. Sometimes one of them can gobble up a nearby nucleon before disappearing. This notion leads to an estimate of 10^{45} –10^{50} years for the elimination of matter as we know it. The peak of 10^{50} years seems critical. But it is located very far into the future: 10^{50} years is ten thousand billion billion billion billion times the present age of the universe. However, the Hawking process, based on black holes, is not in any sense proven. So it is quite legitimate to think about yet more remote goals beyond 10^{50} years. Even if ordinary matter vanishes, supergalactic, galactic and stellar black holes will remain in the universe.

Hawking, however, has come up with yet another stage, which we will number the fifth. According to him, black holes could evaporate in a type of inverse process. If a pair composed of a particle and an antiparticle is created close to a black hole, the antiparticle could fall into the black hole and, under the right circumstances, reduce its mass. The mass loss is accompanied by the release of radiation, in sharp contradiction to the simplistic view that nothing can ever escape from a black hole. The hole evaporates progressively: very slowly to start with, but it gathers momentum, fizzles faster and faster, and finally disappears in a photon flash. Around 10^{100} years from now all black holes will be flitting away in sporadic flashes giving the empty blackness of space a rare taste of action. In the long run the cosmos is only populated by black dwarfs, neutron stars, planets and radiation in this hypothesis.

The stages the universe could go through beyond this project us tremendously far into the future, with sufficient time elapsing for another quantum process, the 'tunnel' effect to take hold. The first objects to profit from this are stars other than neutron stars. In this sixth stage all atomic nuclei that are not already iron do change to iron, the most stable of all nuclei, by around 10^{500} –10^{1500} years from now. According to this notion the universe goes through an immensely long, mildly radioactive, phase which releases a modest amount of energy.

176 The universe and ourselves

The Tunnel Effect

If you want to cross a mountain range in your car you can drive up one side and down the other. If you don't have enough fuel for this then crossing is impossible. In quantum physics the situation is different: a particle, just like your car or even an astronomical object (star, galaxy,...) is localised in space by a wave function which gives the probability of finding it in some given place or other. If your car is on this side of the mountain then the wave function is stongest there. But the mountain itself is not opaque to these probability waves of quantum physics, just as a wall is not a complete barrier to sound. A tiny fraction of the wave propagates through the mountain; for sure the wave on the other side is incredibly weak, but it is present. So there is a certain, but remarkably tiny, chance that your car can flit to the other side of the mountain without actually climbing it. This experience is like driving through a non-existent tunnel. In classical physics, it is impossible to move gratuitously from a state to another that needs first to consume some energy, even if that energy can be recovered later; we have first to be provided with this energy. In quantum physics we can burrow through mountains, using the tunnel effect, without having to pay for any fuel ahead of time.

Now for the seventh step! The tunnel effect changes iron-rich black dwarfs into neutron stars, releasing huge amounts of energy in the form of intense bursts of neutrinos. This does not happen however until $10^{10^{76}}$ years, a number so huge that ordinary power of ten notation breaks down and we get an index (76) on the index (10). To write this down using standard power of ten notation would give $10^{10000000\cdots}$ not with 76 zeroes in the power, but 10^{76} zeroes! That is as many zeroes as there are protons in all the billions of galaxies ever observed. This is the largest number to emerge from the calculations of far future theorists; Freeman Dyson is its father.

An eighth stage might accompany all of this, although the calculations are very sketchy: the neutron stars can tunnel through by quantum processes into black holes (which are more stable) and these then evaporate to radiation. During this time the planets, asteroids and dust can evaporate conventionally into their constituent particles, because over such long periods their atoms have some chance to separate from their neighbours.

By the time we reach the 'Dyson age' the cosmos is just expanding space, cold and just occasionally populated by a photon, a neutrino, or very rarely, a particle. What happens after that? Is the universe condemned to live only on quantum fluctuations of the vacuum into eternity?

We cannot get help simply by throwing these views to one side as so much audacious garbage. They are finally based on the laws of physics as we understand them today. But many weaknesses mar our cosmic odyssey. However, I do not target the number $10^{10^{76}}$ which emerges from uncertain though plausible calculations. Rather, the weak points are our hazy knowledge of the properties of black holes. As was the case when we looked right to the zero point of the Big Bang, it will be necessary for physicists to come up with a solid unification of quantum theory and general relativity in order to tighten up on these calculations. We have already noted other uncertainties: does the universe expand or contract by the time we reach 10^{11} years; are nucleons stable or unstable beyond 10^{32} years... The daring scenarios I have presented are the first forays made by the human mind to so far into the future. So it is quite permissible to say at least that everything is very tentative.

It is bound to remain tentative since everything is based on an appreciation of today's laws of physics. Without question further progress is going to modify those views. I sense that the least predictable factor which could affect the future of the universe is direct action by intelligent beings. The evolution of life on Earth has been so brisk and so varied that it is impossible to have the vaguest idea as to what things will be like in a billion years from now; physics as such cannot help us here. And a billion years is like a brief instant when we set it against the beckoning eons of the far future.

In principle nothing known to science will prevent intelligent life lasting for a very considerable time into the future. The first big catastrophe is scheduled for a billion years hence when the Earth's rotation axis will topple over. By then, technological ways round the problem should have been found. The next challenge on the horizon is ten billion years away when the Sun ceases its career. Then it will be necessary to leave the solar system and find a more hospitable star or stars and take advantage of their nuclear energy until the end of the second stage, about 10^{14} years away.

By degrees, star hopping in essence, it will be necessary to get close to the galactic centre and so avoid being hurled into the hopeless environment of intergalactic space by the gravitational slingshot. This will need to be accomplished by the end of the third stage described above, that is by 10^{18} to 10^{24} years. Life meanwhile would just have to keep going using artificial nuclear energy.

From that stage it will be essential to get into orbit round a nice big black hole, such as the galactic one at the centre of the Galaxy or, better still, the supermassive black hole in the Virgo cluster of galaxies. By orbiting a black hole it will be possible to use its rotation energy for subsistence. If the energy resources there are big enough, life could last until the black hole evaporates, in 10^{100} years, at the end of the fifth stage. That of course is only feasible if nucleon decay has not blotted out all of life's chances during the fourth stage 10^{32} to 10^{50} years.

What aspects of physical laws stop us from considering survival until at least 10^{32} years? Up until then energy is available, as also is everyday matter in the form of nucleons. Life, if it is able to plan rationally, will be able to find niches in an *ad hoc* fashion and so reach a very advanced age.

If this is to happen, who can pretend that it will not have time to develop helpful technologies at present unknown to us. Nobody. Note, however, at this point I am going to engage in unfettered speculation. Encouragingly, Freeman Dyson has already suggested that 'non-biological' forms of life could, in principle, last forever in a universe expanding into eternity. As the temperature of the universe gradually falls, the activity of these life forms would have to decrease as well. A sufficiently resourceful community should be able to gain unlimited longevity by manipulating the environment. It would be necessary for such a community to 'hibernate' for long periods; however the number of active periods would be unlimited, and the total life span of the community could be infinite. Dyson speculates that a limited amount of energy is all that is needed to keep going indefinitely. This last point is at the nub of his thinking and it opens up the possibility of a kind of 'immortality'.

But what purpose would this immortality serve if memory itself is limited? If each of us lived for a thousand years without remembering his first century, in what way would it differ from our present system of parent, child, grandchild,...,and so on? Dyson has, interestingly, described memory systems that have no limits in principle to their capacity, just as the expansion of the universe has no limit. He has also shown that future civilisations could communicate with each other for an indefinite period by using electromagnetic waves. This opens a perspective on activity of unlimited richness, where there will always be something new to discover. Again, quoting Dyson 'I have found a universe of unlimited richness and complexity, a universe in which life can continue indefinitely and can reach neighbours across unimaginable vistas of space and time... There are valid scientific reasons for seriously considering the possibility that life and intelligence could succeed in modifying the universe for its own ends.'

It is heartening to speculate that the murky, erratic, enigmatic, universe of ours might one day find itself, at last, in the certain hand of a different intelligence.

Epilogue

Where has our exploration of the enigma of the universe left us, after all these pages? Clearly, we have encountered facts that are profoundly significant. In terms of spatial dimensions, we have probed the cosmos from the Planck length to quasars and the cosmological horizon. In time we have gone from the Planck time to the Dyson age. We have looked at structures ranging from the trio of quarks lost inside a proton, like three viruses in a volume the size of the Sun, through to the filamentary structure of the universe at large. Our story has embraced particles as subtle as the neutrino and as hypothetical as massive and destructive magnetic monopoles. The recession of the galaxies and the cosmic microwave background are relics of its explosive beginning. Using relativity, with its fusion of space and time, we can show how this Big Bang leads to what I term the grandiose fresco, or the golden moment that started the universe as we know it. As it aged from one second to fifteen billion years, the universe first experienced fifteen minutes of frenetic nuclear activity, followed by a lengthy period of lethargy, lasting a hundred million years when relatively little happened. Then, almost by chance, the stars began to shine, producing the heat and light that gave rise to life on Earth about ten billion years later. This pleasant state of affairs will last about a further hundred billion years. Over such an extended time scale it is impossible to say if greater intelligences will continue the long cosmic odyssey. We can only wonder if the initiatives taken by the International Astronomical Union, through its Commission on Extra-terrestrial Life, will lead to the detection soon of superior civilisations.

Quantum physics gives us surprising insight into the first few moments in the life of the cosmos. In this quantum sea of antiparticles, virtual particles, and so on, we have seen how the electroweak unification led to the prediction of the W and Z bosons, whose existence was brilliantly confirmed at CERN. From this advance, Grand Unification and its fertile application to cosmology, has given a vivid scenario of the interval 10^{-35} to 10^{-32} seconds, when space expanded at a prodigious rate, propelled by the energy of the false vacuum, in which every part of space is charged with 'dynamite', as it were. This phase terminated when the Second Detonation took place, which produced both matter and antimattter, fortunately (miraculously!) slightly favouring matter, so that the cosmos had the material building blocks for the rest of its history. The processing of this material during the next, longest, second, brought the early universe to the onset of the grandiose fresco.

A quite exceptional set of conditions led to the appearance of humans: the specific properties of the nuclei of carbon atoms,, and the presence of liquid water on Earth, in marked contrast to the catastrophes that have taken place on Mars and Venus. Is our universe a uniquely favoured among the myriads of other universes suggested by the 'crystal' analogy we used in the description of inflation?

And what will happen long after today? Contraction or unrestricted expansion? The climax of a hundred billion years of evolution will be reached. If time itself runs on without limit, the stars become spread out, dispersed through progressively larger and emptier volumes of intergalactic space, or they coalesce into gigantic black holes. Any civilisations wishing to escape this face a long exodus, and will end their journey in the vicinity of supergalactic black holes. Dyson suggests a universe of limitless richness and complexity is physically possible.

The ancient enigma of the origin and purpose of the universe, which initially seems a finite problem, has become harder and harder to express simply. If we are to know and to comprehend the universe we first have to jettison a lot of baggage: dogmas, common sense, pre-conceived notions, and emotional or anthopomorphic sentiments. We must then look at everything afresh. Furthermore, the game of the universe set out now does not exclude mankind and any extra-terrestrial cousins from gathering the trump cards, to become masters of the universe, shaping it to their desires and dreams. Of course, the chances of this are minimal, but they are not zero. Are our desires so extravagant? Perhaps not. Consider this verse, from fourteenth century Japan:

> *I, who can cross*
> *a vanishing world*
> *like the foam of the waves,*
> *desire a little fishing boat*
> *more than anything*